Manufacturing Inventory and Supply Analysis

Manufacturing Inventory and Supply Analysis

A Mathematical Modelling Approach

Sanjay Sharma

CRC Press
Taylor & Francis Group
Boca Raton London New York

CRC Press is an imprint of the
Taylor & Francis Group, an **informa** business

First edition published 2022
by CRC Press
6000 Broken Sound Parkway NW, Suite 300, Boca Raton, FL 33487–2742

and by CRC Press

2 Park Square, Milton Park, Abingdon, Oxon, OX14 4RN

CRC Press is an imprint of Taylor & Francis Group, LLC

ISBN: 978-1-032-08170-0 (hbk)
ISBN: 978-1-032-10166-8 (pbk)
ISBN: 978-1-003-21399-4 (ebk)

DOI: 10.1201/9781003213994

Typeset in Times
by Apex CoVantage, LLC

Contents

Preface

A manufacturing inventory system does not only deal with the production of final products, but also the purchase of raw materials or input items. The focus of the present book is on the modelling of such systems incorporating real-world situations such as quality level, among other aspects. The total related cost is optimized after the concerned problem description and formulation. In addition to a conventional inventory system, the described models are also helpful in the business environment such as JIT, MRP and supply chains. With the inclusion of advancements in this area, it will also be beneficial in generating different projects at UG/PG level. The present work is expected to be a reference source for researchers/practitioners in this area, and also for the senior-level students of allied engineering and management disciplines.

Sanjay Sharma
Mumbai, India

About the Author

Sanjay Sharma is a Professor at the National Institute of Industrial Engineering (NITIE), Mumbai, India. He is an operations and supply chain management educator and researcher. He has more than three decades of experience, including industrial, managerial, teaching/training, consultancy and research; he also has many awards/honours to his credit. He has published eight books, and papers in various journals such as the *European Journal of Operational Research*, the *International Journal of Production Economics*, *Computers & Operations Research*, the *International Journal of Advanced Manufacturing Technology*, the *Journal of the Operational Research Society* and *Computers and Industrial Engineering.* He is also on the editorial board of a few journals, including the *International Journal of Logistics Management.*

1 Introduction

A manufacturing inventory system mainly consists of: (a) procurement of raw materials and components, and (b) manufacture of components, sub-assemblies and finished products. Inspection and quality control of the raw materials, components and finished goods also play a role in proper understanding of the manufacturing inventory system. The present work is focused on the mathematical modelling of such systems, primarily on procurement and production of items.

The present book is organized in the following chapters:

1. Introduction
2. Batch size
3. Batch size relevance for JIT
4. Price variation
5. Batch size for MRP
6. Multiproduct manufacturing
7. Manufacturing rate flexibility

Industrial or business organizations operate in the variety of systems with wide fluctuations from time to time. Challenges before the managers, as well as professionals, will be to deal with them effectively with some rationale in decision making. Formulation of the problem is a useful tool and often a prerequisite for the solution of different business complexities. In order to undertake in-depth studies, a clear idea of various functions of an industrial organization is essential. To serve this purpose, an industrial structure is discussed briefly in Section 1.1. This includes important functional divisions of an industry such as:

1. Production
2. Inspection and quality control
3. Marketing
4. Finance
5. Accounts

Their implication for production-inventory system optimization is explained since each views inventory in a different context.

Classical categories of production system i.e. job, batch and mass production are described in Section 1.2 for creating a suitable background. Further, just-in-time (JIT) environment and material requirements planning (MRP) systems are discussed due to their significance from application point of view. Two types of costs are significant. These are the costs, including:

1. Those pertaining to ordering of any material/components
2. Setting up of a machine for manufacture of any product

DOI: 10.1201/9781003213994-1

These two types of costs are related to different environments, but are mathematically similar. In addition these, inventory carrying costs are included in the models. Depending on the situation, shortage costs may also be incorporated in the inventory system. Shortages may result into either:

1. Complete backordering, or
2. Partial backordering

Partial backordering is also discussed briefly, as it will help in the modelling process to optimize the procurement or production lot size.

Mathematical models have been developed in Chapter 2 by incorporating different types of realistic features. In addition to ordering/setup costs and inventory holding costs, these features include:

1. Purchase/production costs
2. Quality defects
3. Shortages which are completely backordered
4. Partial backordering

An integrated production-inventory model is formulated considering partial backordering and quality defects, apart from various costs.

The concept of 'nil inventory' seems to be very appealing, but this approach is not always economically or practically feasible. However, firms should try to move closer to this goal. Relevance of lot sizing is always there, whether it is JIT or traditional approach. These issues, discussed in Chapter 3, are consistent with the JIT concept. The models are formulated which are also useful in meeting supply chain goals. Production of end items is considered, along with the replenishment of input items. The following aspects are covered:

1. Instantaneous procurement
2. Finite replenishment rate of input item
3. Multiple input items
4. Frequent delivery of produced item

Price discounts are often declared by the manufacturing or supplying firm in order to reduce their inventories or increase their market share. The purchasing firms try to take advantage of these situations in order to decrease their overall procurement costs, but it depends on the present stock position among other features of that particular inventory item at which price discounts are offered. Chapter 4 is concerned with the modelling procedures related to price discounts for replenishment with variety of stock positions, such as:

1. Positive stock status
2. Negative stock status
3. Negative stock equivalent to optimal maximum shortage quantity

Discussion is also presented for a declared price increase. This situation is also analyzed concerning the replenishment start at a variety of stock positions, as mentioned previously.

Various MRP lot sizing procedures have been explained in Chapter 5:

1. Economic ordering quantity
2. Period order quantity
3. Silver–Meal heuristic
4. Least total cost
5. Least unit cost
6. Lot for lot
7. Fixed period requirements

Shortages are generally not considered in an MRP environment. Although shortages are undesirable, as they result in loss of goodwill, these are sometimes unavoidable in the real world. The kind of insight provided in Chapter 5 can be useful for tradeoffs whenever it is not possible to avoid backordering due to one or another problem being faced by an organization. Another heuristic procedure is also introduced which has wide applications, as it does not allow shortages in the MRP environment and an important objective of MRP is achieved. Emphasis is on using it, if it gives lower costs in comparison with the other practical lot sizing rules. It can also be used effectively for the optimization of multilevel production-inventory systems.

Multi-item manufacturing environment is discussed in Chapter 6. Models are presented with backorders, as well as fractional backlogging. Shelf life of a product needs to be considered, because it affects certain production situation, and thus, the shelf life constraint is incorporated. This is related to the use/consumption of an item before a certain specified period known to the management. The following approaches are explained to evaluate the optimal results:

1. Common cycle time
2. Different cycle time

Input item procurement is also incorporated, along with multi-item production scenarios. Flexibility in the production rate is described briefly in the context of a family production environment.

It is always useful to have flexibility in operational systems. Production rate variation is sometimes necessary because of demand variation, among other reasons. In addition to the demand variation, a flexible production rate is analyzed in Chapter 7, along with upward and downward variations. Situations with and without shortages are incorporated, along with the total relevant costs. Upper and lower bounds have been obtained for facility setup costs, in order to ensure an economical variation concerning:

1. Upward demand variation
2. Downward demand variation

3. Upward production rate variation
4. Downward production rate variation

The manuscript provides the detailed models/analysis pertaining to various cases which are useful for MRP, JIT and supply chain environment, as well as traditional production-inventory systems. Interaction effects of various operational parameters on the relevant costs are discussed, along with the managerial implications wherever these are appropriate.

1.1 INDUSTRIAL STRUCTURE

An industrial organization is aimed at the conversion of raw materials into finished products. These products will be consumed by another industry or will be available in the open market through a distributing network. Figure 1.1 shows the industrial

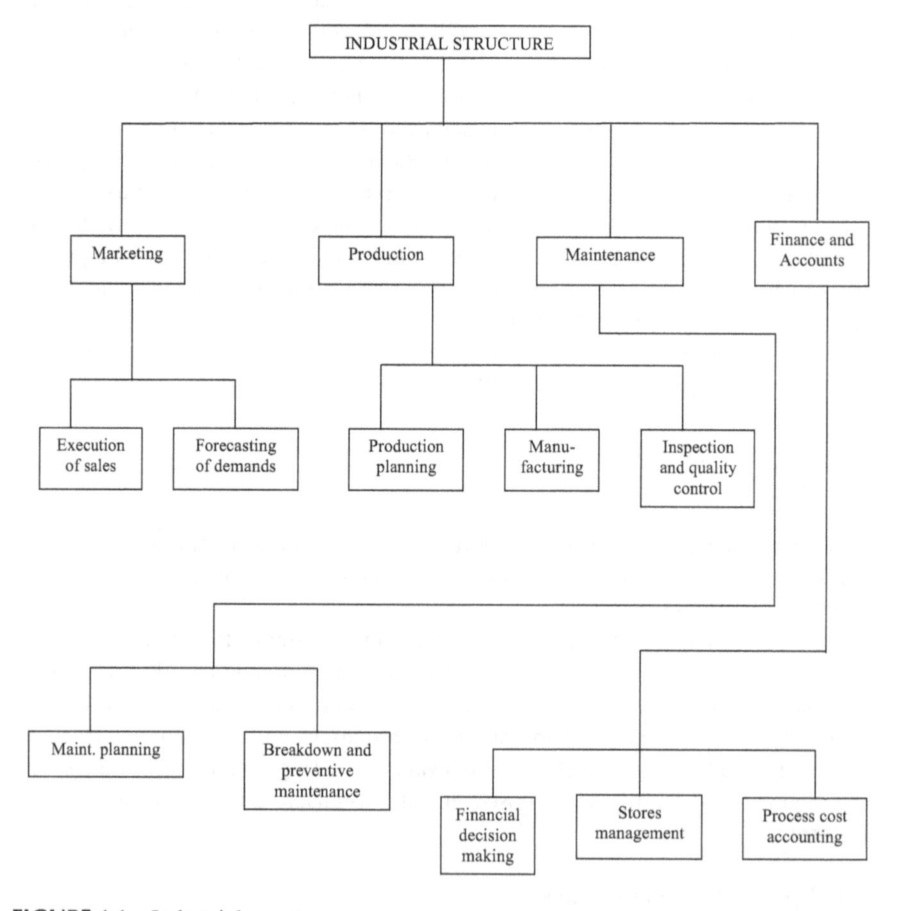

FIGURE 1.1 Industrial structure.

structure with major functions. The marketing division of the industrial organization is engaged in executing the sales of the products, as well as in forecasting the demands. The forecasts made will be useful in generating the production plans, which are targeted by the production division with the assistance of inspection and quality control divisions. For upkeep of the equipment and monitoring the health of the machines, the maintenance division will come into the picture. Its objective is to conduct routine and breakdown maintenance, as well as cost-effective maintenance planning, including spare parts management. However, this is not included in the following sections, since it may not be a significant part of any production-inventory system. Finance and accounts in the present context mainly deal with the capital budgeting and process cost accounting, respectively, in addition to numerous other tasks.

1.1.1 Production

Production is concerned with manufacturing a specified quality product by utilizing available resources in an efficient manner. The efforts of production personnel are directed at achieving the production targets by way of conventional tools—namely, routing, scheduling, dispatching and follow-up. In addition to production control, production planning is a very important task which considers the demand forecasts in input form, and the output is generated in the form of master production schedule (MPS), discussed in Section 1.2.4. MPS is prepared by making use of efficient lot sizing procedures and capacity planning. This is shown in Figure 1.2.

Lot sizing procedures are based on the total cost minimization, whereby total cost usually consists of setup costs and inventory carrying costs which are in conflict with each other.

In the present industrial structure, 'Inspection and quality control' is included as a subdivision of 'Production' to show the strong relationship with reference to the total quality management concept. Even if they are separate functional divisions in many organizations, 'production' and 'quality control' go hand in hand.

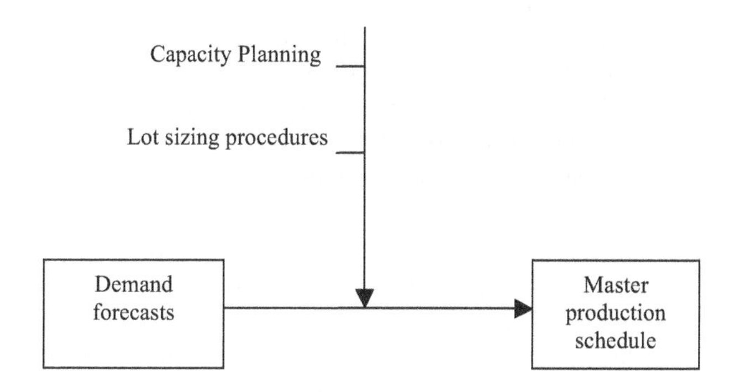

FIGURE 1.2 Generation of MPS.

1.1.2 Inspection and Quality Control

Inspection of any component or product is associated with the detection of defects, whereas quality control is the prevention of defects from occurring. Ignoring or covering up a problem is the worst job one can do. When a worker hesitates in reporting the mistake or malfunction, he keeps on repeating the same, hoping that it will be unnoticed. Workers should be encouraged to identify and report such aspects to their superiors because the worker is usually the first person to notice. It is needless to say that mere inspection does not ensure improvement in the quality of the product. In fact, quality control cannot be viewed in isolation; it is the result of consolidated efforts to improve the processes, maintenance of the machines, etc. It is required for the people to share their experiences, support one another and build commitment together.

Quality level reflects the realism and manufacturing complexities encountered. It is also a major policy decision. For instance, if there is sudden increase in demand for a particular part, the company may like to take advantage of the situation for profit maximizing. It can decide in favour of using an old facility for production of that part, even if it produces more defective parts than its current level. It is of interest to determine the variation of optimum production lot size or inventory ordering policies with respect to quality levels. Mathematical models discussed in the manuscript make use of the proportion of nondefective items in a lot as one of the parameters.

Whenever any lot of raw materials or purchased components arrives in the manufacturing firm, then usually a senior employee of the company visits the site for overall inspection of the lot. He also decides the individual unit inspection procedure and instructs the subordinate staff accordingly. Fixed inspection cost is determined by estimating the costs involved in all such activities. Individual unit inspection cost also needs to be determined, precisely as it finds application in the integrated production-inventory model.

1.1.3 Marketing

An important element of the industrial structure is marketing, because the very survival of the organization depends on it. Major functions of the marketing include sales promotion and providing input data for demand forecasting. There is also close coordination among marketing, production and quality control divisions. This is because complaints received from the customers by the marketing division are usually referred to quality personnel for further investigation and subsequent improvement of the product with the help of manufacturing professionals. This interrelationship is shown in Figure 1.3.

As discussed before, different elements of industrial structure view inventories in different contexts. The marketing division is more concerned with customer satisfaction and supply of products as and when desired by the customers. Therefore, it wants to have a large inventory of finished goods, as well as spare parts. Similarly, the production division desires to have huge inventory of raw materials so that production processes do not suffer for want of material or in case of unforeseen circumstances, including shortages. On the other hand, finance professionals' efficiency lies in the

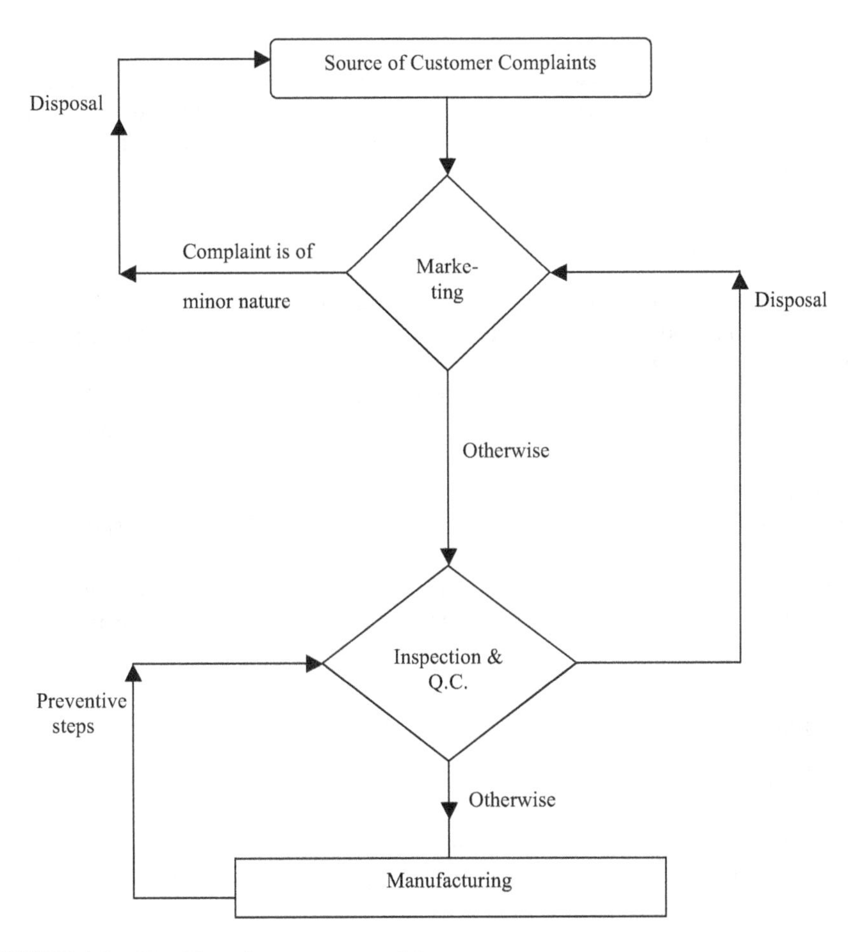

FIGURE 1.3 Handling of customer complaints.

profitable use of a limited budget. They view inventories as a burden because capital is blocked for a longer period. The problem is to optimize the conflicting objectives.

1.1.4 Finance

Capital budgeting is one of the significant tasks of financial management. A large number of proposals for investment usually compete for approval, and there are financial constraints from the time immemorial. The potential proposals/projects are ranked on the basis of profitability. These are arranged in decreasing order of their measure of profitability. In other words, these are sequenced in the order of preference and projects are undertaken depending on the limited capital budget.

Working capital management is another important functional area. Raw material inventory and purchased components form a vital segment of working capital. Analysis of the conflicting costs—such as those of ordering and inventory carrying, decisions regarding how much quantity is to be procured and at what frequency—are to be taken.

1.1.5 Accounts

Apart from the routine tasks of record keeping and updating, etc., process cost accounting and product cost estimation are relevant functions concerning production-inventory planning.

As the raw material progresses towards finishing stage of production, value addition is there at every stage of manufacturing. Though process cost accounting is an essential step for cost optimization, it can also help in estimating the value addition which, in turn, in useful for work-in-process inventory valuation.

In the present section, some of the important functions of the industrial organizations have been explained briefly. Optimum decisions are not possible without integration of various considerations. In view of this, the overview of industrial structure will be useful in the modelling process of a production-inventory system.

1.2 PRODUCTION-INVENTORY SYSTEM

A production-inventory system includes all kinds of inventories pertaining to:

1. Raw materials and purchased components (in-house, as well as in-transit)
2. Work-in-process, involving processed components, sub-assemblies, etc.
3. Finished goods

It is of relevance to have an introduction of various types of production systems.

1.2.1 Background

A manufacturing organization is engaged in the conversion of raw materials into components, sub-assemblies and finally, finished products, by use of resources. These resources include plants and machinery, human resources, materials and capital. Before setting up of an industry, a lot of planning is required. This includes forecasting of demand, site selection and type of plant layout suitable for producing the demand.

Product life cycle can also help in demand forecasting, as well as production planning. Whenever any product is introduced in the market, the demands are initially low. Then demands grow at a faster rate if the product is appreciated by the consumers. Now the product is at maturity where there is peak demand, after which decline is bound to occur. Before decrease in demand occurs, the promoters must think about diversification and modifications.

Though the concept of product life cycle was developed for tangible products, it is well applicable for concepts and services. This is also relevant for the IT sector, where life cycles are still shorter. Detailed planning should be in tune with the assessment of demand data. To confine the discussion to the manufacturing industry, one must consider job shop, batch or mass production. All three kinds of production systems are explained in the next section.

1.2.2 Job Shop, Batch and Mass Production

In case of items whose demands are too low or the standardization of product design is not possible because customers demand a variety of product designs, job shop production is said to occur. The production facility is devoted to the wide variety of jobs, which is the reason for the name 'job shop'. A job shop may be classified as either an —'open job shop' or a 'closed job shop'.

An example of open job shop is the roadside machine shop which is open to all sorts of customers. Forecasting is very difficult because it is not easy to say what kind of jobs will arrive and when. On the other hand, forecasting up to some extent may be made in case of a closed job shop because this caters to the needs of a particular organization. A closed job shop is usually in the premises of an organization and does not entertain the outside customers.

Batch production deals with demands which are neither too low nor very high. The production rate is more than the demand rate and the facilities usually wait for demands to accumulate because it is economical to produce in a batch. In case of batch as well as job shop production, flexibility is required because variety of products is greater. This is the reason why a process or functional type of layout is used in both kinds of production. Similar types of machines are grouped together in such layout.

Mass production is suitable when demands are higher than the production rate and it is advisable to produce continuously. This is also called continuous production. Product variety is much less. Sometimes even a single type of standard product is manufactured. Product or line layout is appropriate in which the machines are arranged in sequence according to the operations to be conducted one after another. The layout is not flexible, but it is efficient in production. Lot sizing is an important feature, whether it is mass or batch production. For example, in the case of tube industry, though the production is continuous in nature and only tubes or pipes will be manufactured, these are in different sizes. For variety of sizes, different settings are required. Since the setup cost is a significant parameter of the total cost, one has to make a decision regarding the lot size.

1.2.3 Just-in-Time (JIT) Environment

Let us consider the following situations:

1. Production process is perfect in the sense that not a single defective piece is produced. The personnel are trained and experienced enough to deliver quality product without any waste. Scraps and rework are completely avoided.

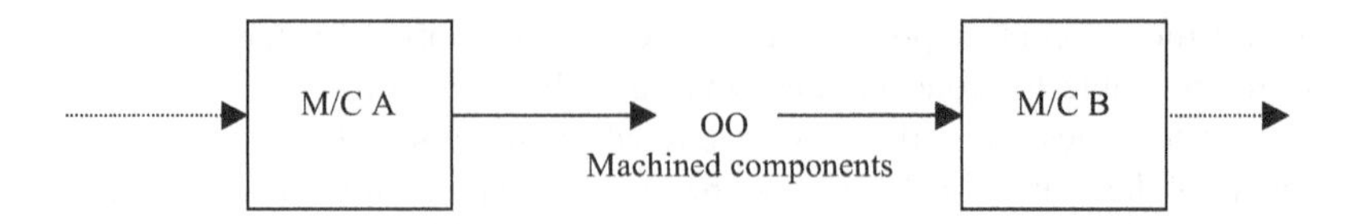

FIGURE 1.4 Flow of material in production shop.

2. Preventive maintenance procedures are implemented effectively, so that no breakdown of machine occurs.
3. Raw materials and purchased components arrive in the plant whenever they are needed, i.e. just-in-time.
4. Lead time is negligible, i.e. items are produced whenever desired. In other words, the setup time for the machine is very short, or almost negligible. The ideal lot size is one production item.

All these situations may look unrealistic and impossible to achieve. But a just-in-time (JIT) production system is based on such ideal conditions. In practice, the emphasis should be on the smaller lot sizes, reduced lead time and setup time, and almost defect-free production or high proportion of nondefective items in a lot. Ideally, JIT environment should approach to 'zero inventory'; yet practically speaking, inventories must be kept as low as possible. If a high level of inventory is there, it may hide inefficiencies in the utilization of resources. For example, if work-in-process inventory is greater, then it covers the improper use of equipment or lack of process control. This is explained in Figure 1.4, which shows the flow of material on the shop floor.

A and B are two machines among many on the shop floor and machined components which are output of A are shown by sign 'OO'. These components are lying in the intermediate space between two machines and are an input to the machine B for further processing. If the components are greater in number, then it serves as a cushion for further processing. Efficiencies on the part of machine B and its operator are hidden, because if machine B is producing more defective parts, then it has more than enough input material for further trial. These malfunctions may not be visible easily due to more work-in-process inventory. Similarly, if the production speed of machine A is less than the optimum for some reason, its operator may not care for it because more components are lying in the intermediate space and machine B will not stop due to lack of production of machine A. On the other hand, if work-in-process inventory is less or the smaller lot sizes are there, the pressure on the resources of all kinds is more to perform optimally, which is beneficial from the management point of view.

Product/Process Cost Estimation as an Aid

Lower inventory levels are characteristic of the JIT environment, but relevance of inventory valuation still exists. Product/process cost estimation is useful for the final product, as well as the work-in-process inventory valuation, regardless of whatever kind of production system is adopted. Let us consider a simple example of asbestos cement (A.C.) pressure pipe manufacturing. Cement and asbestos fibres are blended to form a slurry which is admitted into a vat containing a rotating sieve cylinder covered with a fine mesh. This layer of asbestos cement is then passed over a mandrel whose diameter is equivalent to the inner diameter of the A.C. pressure pipe.

Data related to 100mm class 10 A.C. pressure pipe are as follows:

Weight of the pipe = 7.79 kg/m
Weight of the asbestos fibre = Approx. 14% of the pipe weight
= 0.14 × 7.79
= 1.091 kg/m
Remaining constituent is cement and its weight = 7.79 – 1.091
= 6.699 k/m

Consider the purchase costs of asbestos fibre and cement as Rs. 60/kg and Rs. 2/kg, respectively.

Direct material cost = Material cost of (asbestos fibre + cement)
= (1.091 × 60) + (6.699 × 2)
= Rs. 78.86 per m of pipe

If manufacturing cost is estimated to be 15% of the direct material cost, then cost of the manufactured product is:

= 1.15 × 78.86
= Rs. 90.69 per m

This is an illustration of the two-echelon production-inventory system shown in Figure 1.5. However, for complex assembly structure, manufacturing cost at each

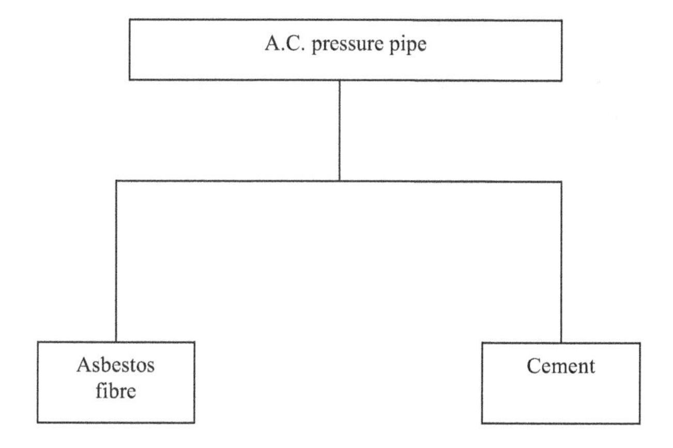

FIGURE 1.5 Inventory structure for asbestos cement pressure pipe.

stage should be estimated and value additions should be made as per the stages of production. This will be based on the individual process cost estimation. The product/process cost estimation is an essential exercise to facilitate the end product and work-in-process inventory valuation.

1.2.4 Material Requirements Planning (MRP) System

Consider that a finished product A requires two components: B and C as shown in Figure 1.6. Assume that the master production schedule (MPS) requires 15 as the number of finished product A each in week Nos. 3 and 4 such as,

Week No.:	**1**	**2**	**3**	**4**
Quantity:	0	0	15	15

Component B is produced inside the factory premises, whereas component C is procured from outside sources. Lead time for manufacturing of B is estimated to be two weeks, and lead time for procurement of C is one week.

Given the MPS, MRP (material requirements planning) is used to plan the manufacturing or purchase of the components. Quantity of the component required per unit finished product is also needed to generate the schedule. If one unit of component B is required per unit finished product A, then the planned production of B, considering lead time of two weeks, is as follows:

Week No.:	**1**	**2**	**3**	**4**
Quantity:	15	15	0	0

If two units of component C are needed per unit production of A, then the planning for procurement of C, considering lead time of one week, is as follows:

Week No.:	**1**	**2**	**3**	**4**
Quantity:	0	30	30	0

Similar logic can be applied for a general product structure (shown in Figure 1.7) by exploding the MPS with the use of bill of material and lead time.

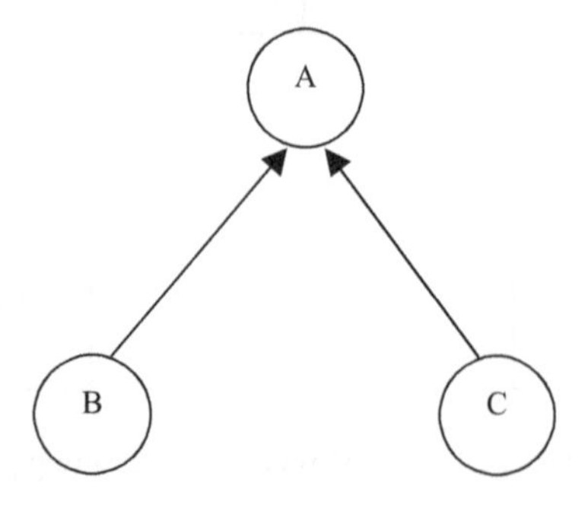

FIGURE 1.6 Product structure.

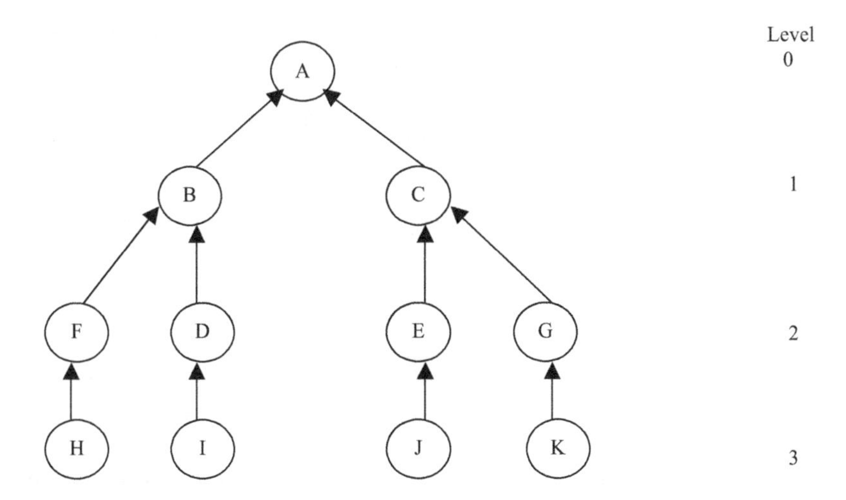

FIGURE 1.7 General product structure.

At Level 0, there is the finished product or final assembly A. B and C are the sub-assemblies at Level 1. Components D and F are used to produce subassembly B. Similarly, components E and G are needed for the subassembly C. Components D, E, F and G are shown at Level 2. Raw material or input item H is required to manufacture component F. Similarly, raw materials I, J and K are needed to produce the components D, E and G, respectively. Raw materials H, I, J and K are shown in Figure 1.7 at the lowest level, i.e. Level 3.

As the components D and F constitute the subassembly B, the planned requirements of B is used as an input for generating the requirements of D and F. In this process of explosion, the quantity of D and F required per unit of B will be utilized. In order to generate the MRP schedule, lead time is used as explained before, either for production or for procurement of items.

Exercises

1. Explain a production-inventory system.
2. Discuss various divisions and subdivisions of industrial structure.
3. Consider only the following small portion of industrial structure. Describe the interaction between 'marketing' and 'production'.

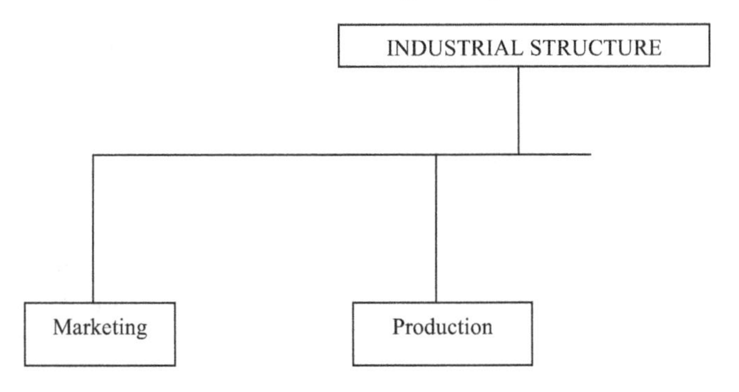

4. Describe the interrelationship of production and maintenance, considering the relevant portion of industrial structure as follows:

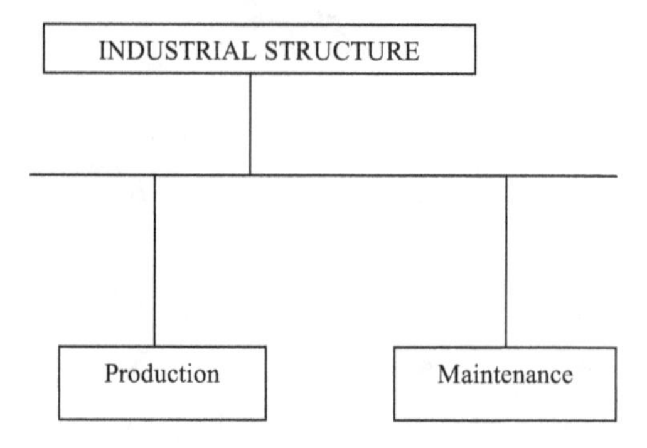

5. In the following figure, 'Production' and 'Finance and accounts' are shown as certain portions of complete structure. Discuss the effect of interaction among these functions.

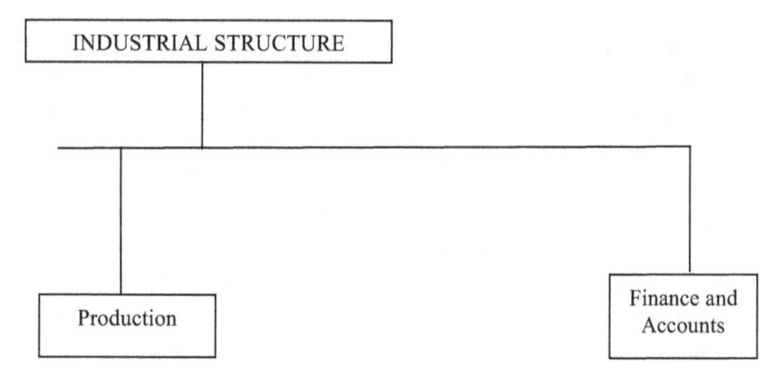

6. As shown following, after forecasting the demand, a company may approach a master production schedule. In this approach, what are various steps?

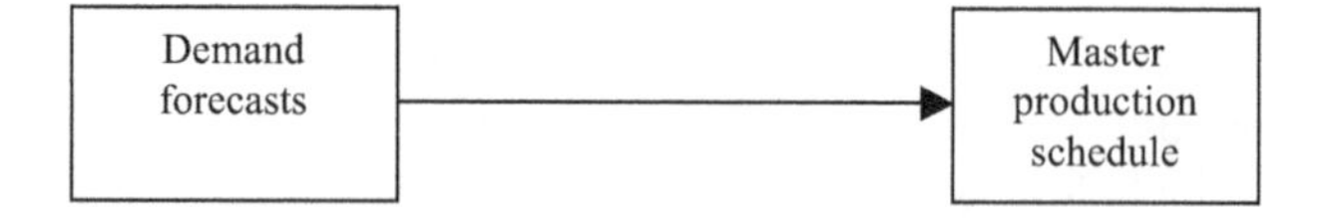

7. What do you understand by 'inspection' and 'quality control'?
8. Discuss the following:

   a. Quality level
   b. Lot
   c. Defective and nondefective items
   d. Proportion of nondefective items in a lot

9. Minor customer complaints might be handled by the marketing division, as shown in the following figure. However, if this is not the case, describe

the complaint handling procedure in detail, including other divisions/sub-divisions in an industrial organization.

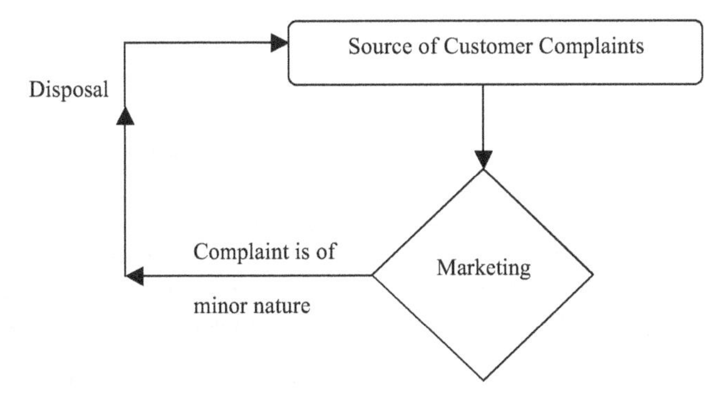

10. Discuss the role of 'Finance and accounts', particularly in the context of inventories.
11. Describe the following:

    a Open job shop
    b. Closed job shop
    c Batch production
    d. Mass production

12. What do you understand to be the meaning of JIT environment?
13. Explain various types of problems/situations that might appear in the flow of material in a production shop as shown following:

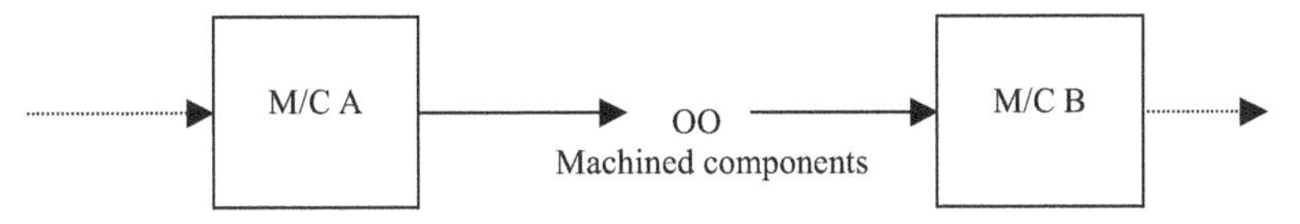

14. Explain the following statement with suitable example: "The product/process cost estimation is an aid to facilitate the end product and work-in-process inventory valuation."
15. Consider that a finished product A requires two components B and C, as shown following:

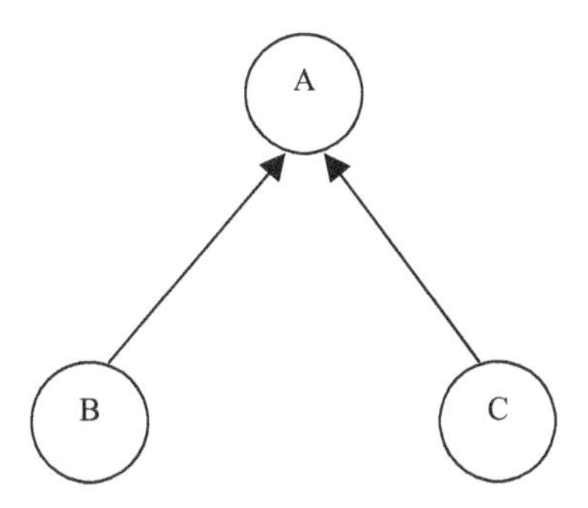

Assume that the master production schedule (MPS) requires 20 of finished product A each in week Nos. 3 and 4 such as:

Week No.:	1	2	3	4
Quantity:	0	0	20	20

Component B is procured from outside sources, whereas component C is produced inside the factory premises. Lead time for procurement of B is estimated to be one week, and lead time for production of C is two weeks. Given the MPS, MRP (material requirements planning) is used to plan the purchase or manufacturing of the components. Quantity of the component required per unit finished product is also needed to generate the schedule.

a. If one unit of component B is required per unit finished product A, then show the procurement plan of B.
b. If two units of component C are needed per unit production of A, then show the planning for production of C.

16. Explain a general product structure with the help of Fig.

2 Batch Size

A company ABC Ltd. is involved in manufacturing of certain product. In order to manufacture it, some components are produced in-house and some components are purchased from different suppliers. It is observed that total cost concerning the purchase of an important component is very high. The organization is making an effort to minimize the total relevant cost. Economic lot size is the quantity to be purchased/produced so that total costs are minimum.

2.1 OPTIMAL PURCHASE QUANTITY

Let:

Annual demand = R
Demand rate per period = r
Proportion of nondefective items in a lot = y
Ordering quantity = Q
Ordering cost = C
Unit purchase cost = P
Annual inventory carrying cost fraction = F

The consumption pattern of the component is shown in Figure 2.1.

Quantity Q, which includes defective items, is being ordered during each cycle. The proportion of nondefective items in each lot of quantity Q is estimated to be y. Each item in the lot is inspected and defective items are rejected and are not added in the inventory.

As the annual demand for acceptable components is R, total quantity to be purchased is (R/y) annually and annual purchase cost = $\frac{R}{y} \cdot P$

$$\text{Number of orders placed in one year} = \frac{R}{yQ}$$

$$\text{Annual ordering cost} = \frac{RC}{yQ}$$

Cycle time (C.T. in periods) = $\frac{yQ}{r}$ and cycle time in years = $\frac{yQ}{r} \cdot \frac{r}{R} = \frac{yQ}{R}$

$$\text{Annual inventory carrying cost per unit} = PF$$

$$\text{Average inventory during the cycle} = \frac{yQ}{2}$$

$$\text{Annual inventory carrying cost} = \frac{yQ}{2}\left(\frac{yQ}{R}\right)\left(\frac{R}{yQ}\right)PF$$

DOI: 10.1201/9781003213994-2

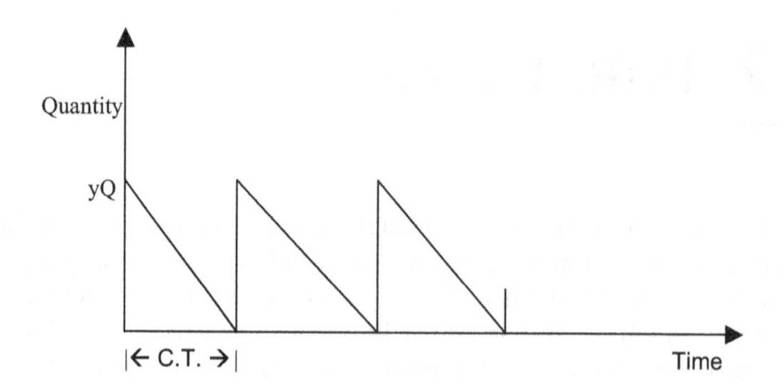

FIGURE 2.1 Consumption with constant demand rate.

as there are $\frac{R}{yQ}$ cycles in one year.

and therefore annual carrying cost $= \frac{yQ}{2} \cdot PF$

[Note: As the average inventory throughout the year is yQ/2, annual carrying cost can simply be said as (yQ/2) × PF. But in more complex models when inventory exists for a fraction of cycle time along with numerous other parameters, this analytical approach is useful for evaluating annual carrying cost.]

Now the total relevant cost is the sum of annual purchase, ordering and carrying cost, and therefore the total annual cost, $E = \frac{R}{y} \cdot P + \frac{RC}{yQ} + \frac{yQ}{2} \cdot PF$ (2.1)

$\frac{\partial E}{\partial Q} = 0$ gives an optimal ordering quantity,

$$Q^* = \sqrt{\frac{2RC}{y^2 PF}} \tag{2.2}$$

If quality defects are not considered, then y = 1 and Equation 2.2 reduces to a well-known economic ordering quantity (EOQ), i.e. $\sqrt{\frac{2RC}{PF}}$

Example 2.1

Consider following data:

Annual demand for nondefective items = 12,000
Demand rate per period (say per month) = 1,000
Fixed ordering cost per order = Rs. 100
Unit purchase cost = Rs. 10

Annual inventory carrying cost fraction = 0.2
Proportion of nondefective items = 0.96

Determine the economic procurement lot size, optimum total annual cost and cycle time in periods.

As R = 12,000, C = Rs. 100, y = 0.96, P = Rs. 10 and F = 0.2, using Equation 2.2,

$$Q^* = \sqrt{\frac{2 \times 12000 \times 100}{0.96^2 \times 10 \times 0.2}}$$
$$= 1{,}141.09 \simeq 1{,}141$$

Substituting this value of Q*, as well as other values in Equation 2.1, optimum total annual cost = Rs. 127,190.89.

$$\text{Cycle time in periods} = \frac{yQ}{r}$$
$$= \frac{0.96 \times 1141}{1000}$$
$$= 1.095 \text{ months}$$

2.1.1 Dealing with Shortages

If shortages are allowed in the inventory system, then machine may remain idle for some time. For a trading firm, shortages may result in back ordering or lost sales. Shortage cost is a parameter to be estimated which will represent such effects. This cost is obviously higher than procurement cost of the item.

Let additional parameters be:

J = maximum shortage quantity
K = shortage cost per unit-year

Inclusion of shortages is shown in Figure 2.2, with negative inventory position represented.

As discussed before, C.T. in years = yQ/R

Shortages exist for time t_2, which is J/r periods

Or $\frac{J}{r} \cdot \frac{r}{R} = \frac{J}{R}$ in years.

Shortage quantity is varying from 0–J linearly and average shortage quantity = J/2 which exists for a fraction of cycle time, i.e. $\frac{(J/R)}{(yQ/R)} = \frac{J}{yQ}$

$$\text{Annual shortage cost} = \frac{J}{2} \cdot \left(\frac{J}{yQ} \right) \cdot K = \frac{J^2 K}{2yQ} \tag{2.3}$$

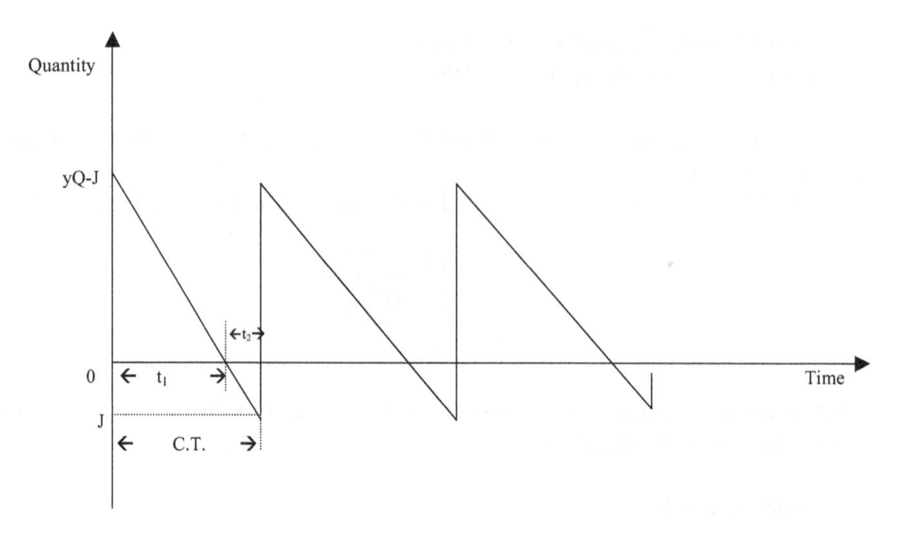

FIGURE 2.2 Consumption with constant demand rate, including shortages.

Inventory exists for time t_1 = C.T. – t_2

$$= \frac{yQ}{R} - \frac{J}{R} = \frac{yQ - J}{R} \text{ in years}$$

Average inventory is (yQ – J)/2 which exists for a fraction of cycle time i.e. $\frac{(yQ-J)/R}{(yQ/R)} = \frac{(yQ-J)}{yQ}$

[Note: In other words, inventory exists for t_1 = (yQ – J)/R year in one cycle and there are (R/yQ) cycles in one year; therefore, the inventory exists for a total time of $\frac{(yQ-J)}{R} \cdot \frac{R}{yQ} = \frac{(yQ-J)}{yQ}$ annually.]

$$\text{Annual inventory carrying cost} = \frac{(yQ-J)}{2} \cdot \frac{(yQ-J)}{yQ} \cdot PF$$

$$= \frac{(yQ-J)^2 PF}{2yQ} \tag{2.4}$$

$$\text{Annual purchase and ordering cost} = \frac{R}{y} \cdot P + \frac{R}{yQ} \cdot C \tag{2.5}$$

Total relevant cost is the sum of Equations 2.3–2.5, given as

$$E = \frac{RP}{y} + \frac{RC}{yQ} + \frac{(yQ-J)^2 PF}{2yQ} + \frac{J^2 K}{2yQ}$$

$$\text{Or } E = \frac{RP}{y} + \frac{RC}{yQ} + \frac{PFyQ}{2} - PFJ + \frac{J^2PF}{2yQ} + \frac{J^2K}{2yQ} \tag{2.6}$$

Now the objective is to determine optimal Q and J.

$$\frac{\partial E}{\partial J} = -PF + \frac{J}{yQ}\left[PF + K\right] = 0$$

$$\text{Or } J = \frac{PFyQ}{(PF + K)} \tag{2.7}$$

$$\text{And } \frac{\partial E}{\partial Q} = \frac{RC}{yQ^2} + \frac{PFy}{2} - \frac{(PF + K)J^2}{2yQ^2} = 0$$

$$\text{Or } Q^2 = \frac{2RC + (PF + K)J^2}{PFy^2}$$

Substituting the value of J from Equation 2.7:

$$Q^2 = \frac{2RC}{PFy^2} + \frac{PFQ^2}{(PF + K)}$$

$$\text{Or } Q^* = \sqrt{\frac{2RC(PF + K)}{KPFy^2}} \tag{2.8}$$

Substituting Equation 2.8 in Equation 2.7:

$$J^* = \sqrt{\frac{2RCPF}{K(PF + K)}} \tag{2.9}$$

As it is a two-variables optimization (minimization) problem, the following conditions need to be satisfied:

$$\frac{\partial^2 E}{\partial Q^2} \cdot \frac{\partial^2 E}{\partial J^2} > \left[\frac{\partial^2 E}{\partial Q \partial J}\right]^2 \tag{2.10}$$

$$\text{and } \frac{\partial^2 E}{\partial Q^2} > 0 \text{ or } \frac{\partial^2 E}{\partial J^2} > 0 \tag{2.11}$$

Now $\frac{\partial^2 E}{\partial Q^2} = \frac{2RC}{yQ^3} + \frac{(PF + K)J^2}{yQ^3}$, which is greater than zero

$$\frac{\partial^2 E}{\partial J^2} = \frac{(PF + K)}{yQ} \text{ and } \frac{\partial^2 E}{\partial Q \partial J} = \frac{-(PF + K)J}{yQ^2}$$

First condition, i.e. Condition 2.10, gives:

$$\left[\frac{2RC}{yQ^3}+\frac{(PF+K)J^2}{yQ^3}\right]\frac{(PF+K)}{yQ} > \frac{(PF+K)^2J^2}{y^2Q^4}$$

$$\text{Or } \frac{2RC(PF+K)}{y^2Q^4}+\frac{(PF+K)^2J^2}{y^2Q^4} > \frac{(PF+K)^2J^2}{y^2Q^4}$$

Which is true, therefore Equation 2.8 and Equation 2.9 give optimal ordering and shortage quantity, respectively. Optimal total annual cost is obtained by substituting Equation 2.8 and Equation 2.9 in Equation 2.6.

Example 2.2

Consider example problem 2.1 in which R = 12,000, C = Rs. 100, y = 0.96, P = Rs. 10 and F = 0.2. Assume additional parameter K (i.e. shortage cost per unit-year) as Rs. 50. Obtain optimal ordering and shortage quantity. Also evaluate the optimum total annual cost.

$$\text{From Equation 2.8, } Q^* = \sqrt{\frac{2\times 12000\times 100(2+50)}{50\times 10\times 0.2\times 0.96^2}}$$

$$= 1{,}163.68 \simeq 1{,}164$$

$$\text{From Equation 2.9, } J^* = \sqrt{\frac{2\times 12000\times 100\times 10\times 0.2}{50(2+50)}}$$

$$= 42.97 \simeq 43$$

From Equation 2.6, optimum total annual cost,

$$E^* = \frac{12000\times 10}{0.96}+\frac{12000\times 100}{0.96X1164}+\frac{10\times 0.2\times 0.96\times 1164}{2}-10\times 0.2\times 43$$

$$+\frac{43^2\times 10\times 0.2}{2\times 0.96\times 1164}+\frac{43^2\times 50}{2\times 0.96\times 1164}$$

$$= \text{Rs. } 127{,}148.34$$

Compare this solution with that of the Example 2.1, in which total optimal cost was obtained as Rs. 127,190.89. By allowing the shortages, the cost is decreased by Rs. 42.55 (127,190.89 – 127,148.34).

This can also be shown analytically as follows:

Substituting Equation 2.2 in Equation 2.1 and on solving,

Total optimal cost without shortages $E_1 = \frac{R}{y}\cdot P+\sqrt{2RCPF}$

Substituting Equation 2.8 and Equation2.9 in Equation 2.6 and on solving, total optimal cost considering shortages,

$$E_2 = \frac{R}{y} \cdot P + \sqrt{\frac{2RCPFK}{(PF+K)}}$$

$$\text{Now } E_1 - E_2 = \sqrt{2RCPF}\left(1 - \sqrt{\frac{K}{(PF+K)}}\right) \tag{2.12}$$

As K/(PF + K) is less than 1, $E_1 - E_2$ is greater than zero. Putting the numerical values,

$$E_1 - E_2 = \sqrt{2 \times 12000 \times 100 \times 10 \times 0.2}\left(1 - \sqrt{\frac{50}{(2+50)}}\right)$$

$$= \text{Rs. } 42.55$$

Hence, a saving of Rs. 42.55 is obtained by allowing shortages. One should not conclude that shortages should be preferred all the time. As the shortage cost increases, savings decrease. This is because of increase in K/(PF + K), with an increase in K, and therefore the factor $\left(1 - \sqrt{K/(PF+K)}\right)$ in Equation 2.12 decreases. For instance, when K = Rs. 100, $E_1 - E_2$ = Rs. 21.59.

Therefore, when shortage costs are very high, savings may not be significant from a management point of view. A decision to allow for shortages also depends on the perception of the management regarding satisfactory service level. Service level is the ratio of demands which are satisfied and total demand.

2.1.2 Partial Backordering

When shortages occur, these may result in backordering or lost sales. Partial backordering, or fractional backordering, refers to the situation when a fraction of shortage quantity is not backordered. At present, a common estimate of shortage cost is associated with all the shortages, whether these result into backordering or not. This estimate of shortage cost may also include activities such as advertising, etc., which are needed to replace the lost customer.

Now, as discussed in Section 2.1.1, annual shortage quantity = $J^2/(2yQ)$, since the average shortages J/2 exist for a fraction of cycle time J/(yQ).

$$\text{Annual shortage cost } = \frac{J^2 K}{2yQ} \tag{2.13}$$

Let:

Fraction of shortage quantity which is not backordered= b

$$\text{Annual quantity which is not backordered } = \frac{bJ^2}{2yQ}$$

$$\text{Annual purchase cost } = \left(\frac{R}{y} - \frac{bJ^2}{2yQ}\right)P \tag{2.14}$$

$$\text{Annual ordering cost} = \left(\frac{R}{y} - \frac{bJ^2}{2yQ}\right)\frac{C}{Q} \tag{2.15}$$

As explained in Section 2.1.1,

$$\text{Annual inventory holding cost} = \frac{(yQ - J)^2 PF}{2yQ} \tag{2.16}$$

Adding Equations 2.13–2.16, total relevant cost:

$$E = \left(\frac{R}{y} - \frac{bJ^2}{2yQ}\right)P + \frac{J^2K}{2yQ} + \left(\frac{R}{y} - \frac{bJ^2}{2yQ}\right)\frac{C}{Q} + \frac{(yQ - J)^2 PF}{2yQ}$$

Or

$$E = \frac{RP}{y} - \frac{bJ^2P}{2yQ} + \frac{J^2K}{2yQ} + \frac{RC}{yQ} - \frac{bJ^2C}{2yQ^2} + \frac{PFyQ}{2} - PFJ + \frac{PFJ^2}{2yQ} \tag{2.17}$$

Equation 2.17 can be minimized by applying search procedures such as univariate method in order to obtain optimal Q and J. Starting point for Q and J may be considered as given by Equation 2.8 and Equation 2.9 when all the shortages were backordered. However, for all practical purposes, very close to optimal solution is obtained by use of the following approximation:

$$\text{Consider Equation 2.15, number of orders} = \frac{R}{yQ} - \frac{bJ^2}{2yQ^2}$$

Second term is $\frac{b}{2y}\left(\frac{J}{Q}\right)^2$, where b and y are fractions.

$J/Q \ll 1$, and therefore $(J/Q)^2$ is a very small fraction; $(bJ^2)/(2yQ^2)$ may be ignored, or the assumption that the frequency of ordering is unchanged with partial backordering is not a serious one, and with this assumption, total relevant cost:

$$E = \frac{RP}{y} - \frac{bJ^2P}{2yQ} + \frac{J^2K}{2yQ} + \frac{RC}{yQ} + \frac{PFyQ}{2} - PFJ + \frac{PFJ^2}{2yQ} \tag{2.18}$$

$\frac{\partial E}{\partial J} = 0$ shows

$$J = \frac{PFyQ}{(K + PF - bP)} \tag{2.19}$$

$\frac{\partial E}{\partial Q} = 0$ shows

$$Q^2 = \frac{2RC + J^2(K + PF - bP)}{PFy^2} \tag{2.20}$$

Substituting the value of J from Equation 2.19,

$$Q^* = \sqrt{\frac{2RC(K + PF - bP)}{PFy^2(K - bP)}} \tag{2.21}$$

Substituting Equation 2.21 in Equation 2.19,

$$J^* = \sqrt{\frac{2RCPF}{(K - bP)(K + PF - bP)}} \tag{2.22}$$

Optimal cost is obtained from Equation 2.18 by substituting optimum Q and J from Equation 2.21 and Equation 2.22, respectively. Conditions which need to be satisfied for two variables optimization problem are given by Equation 2.10 and Equation 2.11.

$$\frac{\partial^2 E}{\partial Q^2} \cdot \frac{\partial^2 E}{\partial J^2} > \left[\frac{\partial^2 E}{\partial Q \partial J} \right]^2 \text{ gives}$$

$$\frac{2RC}{y^2 Q^4}(K + PF - bP) > 0$$

$$\text{and } \frac{\partial^2 E}{\partial J^2} > 0 \text{ gives } \frac{1}{yQ}(K + PF - bP) > 0$$

Both the conditions suggest that K + PF – bP > 0.

As in the real environment, K > P and the condition is easily satisfied.

Example 2.3

Consider Example 2.2 in which R = 12,000, C = Rs. 100, y = 0.96, P = Rs. 10, F = 0.2 and K = Rs. 50.

Assume b = 0.3. Evaluate optimum total cost after obtaining optimal shortage and ordering quantity to deal with partial backordering.

$$\text{From Equation 2.22, } J^* = \sqrt{\frac{2 \times 12000 \times 100 \times 10 \times 0.2}{(50 - 3)(50 + 2 - 3)}}$$
$$= 45.65$$

$$\text{From Equation 2.21, } Q^* = \sqrt{\frac{2 \times 12000 \times 100(50 + 2 - 3)}{10 \times 0.2 \times 0.96^2 (50 - 3)}}$$
$$= 1{,}165.11$$

Substituting values of Q* and J* in Equation 2.18, optimal cost, E = Rs. 127,145.71.

2.2 OPTIMAL PRODUCTION QUANTITY

Batch production circumstances are shown in Figure 2.3.

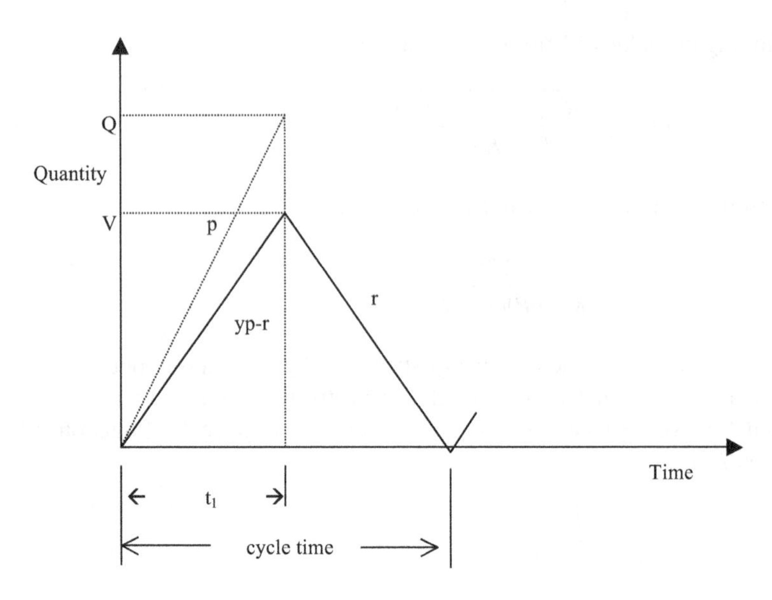

FIGURE 2.3 Production-inventory cycle.

Let p = production rate in units per period.

Production rate is greater than demand rate per period in batch production. Further, y is the proportion of nondefective items in a lot produced by machine, and yp is the effective production rate for acceptable items. Demand is always there during the cycle time, but production time is for t_1 only during the cycle time. In one setup of the machine, quantity Q is being produced at the rate p in units per period, as shown by dotted line in Figure 2.3. But simultaneously, consumption also occurs at demand rate r in units/period. Inventory build-up rate is (yp − r) during production time t_1. For the remaining portion of the cycle time, there is no production, but consumption occurs at rate r until zero quantity is reached. The similar cycle then starts again.

Now, setup cost for machine = C
Production cost per unit = P
Maximum inventory during the cycle = V
Other notation are as before, i.e. annual demand R and annual carrying cost fraction F.

$$\text{As } t_1 = \frac{V}{yp - r} = \frac{Q}{p}, V = \frac{(yp - r)Q}{p}$$

$$\text{Annual carrying cost } = \frac{V}{2} PF$$

$$= \frac{(yp - r)PFQ}{2p} \tag{2.23}$$

As there are R/(yQ) production cycles in a year,

$$\text{Annual setup cost } = \frac{RC}{yQ} \tag{2.24}$$

$$\text{Annual production cost } = \frac{R}{y} \cdot P \tag{2.25}$$

Adding (2.23), (2.24) and (2.25), total annual cost,

$$E = \frac{(yp - r)PFQ}{2p} + \frac{RC}{yQ} + \frac{PP}{y} \tag{2.26}$$

Differentiating partially with respect to Q and equating to zero,

$$\frac{(yp - r)PF}{2p} - \frac{RC}{yQ^2} = 0$$

$$\text{Or } Q^* = \sqrt{\frac{2pRC}{(yp - r)PFy}} \tag{2.27}$$

This is the economic production or batch quantity.
Optimal total cost is obtained by substituting Q* in Equation 2.26.

Example 2.4

Consider annual demand, R = 12,000 units
Setup cost, C = Rs. 200
Proportion of nondefective items in a lot, y = 0.98
Annual inventory carrying cost fraction, F = 0.2
Production cost per unit, P = Rs. 10
Demand rate per period, r = 1,000 units
Production rate per period, p = 1,200 units

$$\text{Using Equation 2.27, } Q^* = \sqrt{\frac{2 \times 1200 \times 12000 \times 200}{(0.98 \times 1200 - 1000) \times 10 \times 0.2 \times 0.98}}$$
$$= 4{,}086.27$$

Substituting in Equation 2.26, total cost is obtained as Rs. 123,647.62

Production time in a cycle, t_1 = Q/p = 4,086.27/1,200 = 3.4 periods (in the present example, months).

In year, t_1 = 3.4 × (r/R) = 3.4x (1/12) = 0.28 year

Cycle time = Q/p + V/r

$$= \frac{Q}{p} + \frac{(yp-r)Q}{pr}, \text{ as } V = \frac{(yp-r)Q}{p}$$

$$= \frac{yQ}{r}$$

In other words, quantity yQ is available for consumption in each cycle and it is consumed at rate r. Therefore the cycle time in periods is $\frac{yQ}{r}$.

$$\text{C.T.} = \frac{0.98X4086.27}{1000} = 4.004 \text{ periods}$$

Or C.T. = 4.004 × (r/R) = 4.004 × (1,000/12,000) = 0.33 year

In ideal conditions when no defective items are produced, then y = 1 and Equation 2.27 reduces to:

$$Q^* = \sqrt{\frac{2pRC}{(p-r)PF}}$$

It is of interest to conduct sensitivity analysis of lot size and cost with respect to situation when y = 1. The company has many similar machines; some of them are new, whereas others remaining are old. These old machines produce a greater number of defective products in an average lot. For the last one year, there is an increase in demand of the item, which is expected to continue for the next two years also. The company is using old machines also to meet the demand, even if these produce a greater number of defective pieces. The determination of optimum lot size and cost is helpful for further analysis. Q* and E* for different values of y are obtained as follows:

For $y = y_1 = 1$, lot size, $Q_1^* = 3,794.73$ and total cost, $E_1^* =$ Rs. 121,264.91

y	% Decrease over y1 (y1−y) × 100	Q* (Units)	% Increase over Q1* (Q*−Q1*) × 100 / Q1*	E* (Rs.)	% Increase over E1* (E*−E1*) × 100 / E1*
0.98	2%	4,086.27	7.68%	123,647.62	1.96%
0.96	4%	4,442.62	17.07%	126,125.46	4.01%
0.94	6%	4,892.46	28.93%	128,703.30	6.13%
0.92	8%	5,486.38	44.58%	131,385.75	8.35%
0.9	10%	6,324.56	66.67%	134,176.61	10.65%

Optimal cost increases almost proportionately with increase in % defectives (i.e. decrease in proportion of nondefective products, y). But optimal production lot size is more sensitive. Its increase is more as compared to increase in defective parts/products.

2.2.1 Allowing Shortages

This situation is shown in Figure 2.4.

Maximum shortage quantity is J, and as p is the production rate per period, yp is the replenishment rate for acceptable products. Shortages occur during time t_1 and t_4, whereas positive inventory exists during t_2 and t_3.

$t_1 = J/(yp - r)$ and $t_4 = J/r$

$$t_1 + t_4 = \frac{J}{(yp-r)} + \frac{J}{r} = \frac{ypJ}{r(yp-r)} \text{ periods}$$

$$= \frac{ypJ}{r(yp-r)} \cdot \frac{r}{R} = \frac{ypJ}{R(yp-r)} \text{ year}$$

As the average shortage quantity is J/2 and there are R/(yQ) number of cycles in a year,

$$\text{Annual shortage cost} = \frac{J}{2}\left(\frac{ypJ}{R(yp-r)}\right) \cdot \frac{R}{yQ} \cdot K$$

Where K is annual shortage cost per unit.

$$\text{Annual shortage cost} = \frac{pJ^2K}{2(yp-r)Q} \tag{2.28}$$

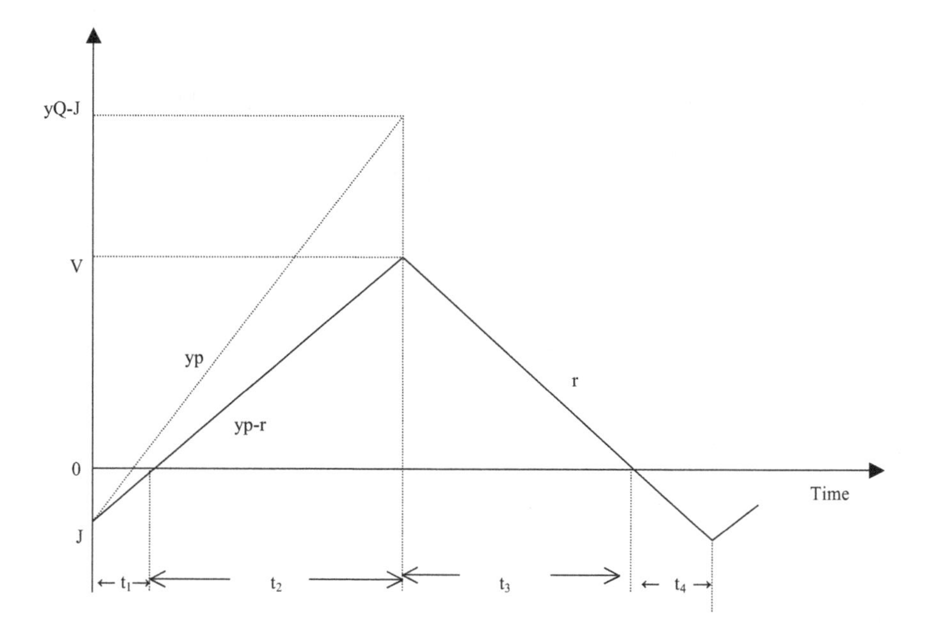

FIGURE 2.4 Allowing shortages in the production system.

$$\text{Now } t_1 + t_2 = \frac{V+J}{(yp-r)} = \frac{yQ}{yp}$$

$$\text{Or } V = (yp-r)\frac{Q}{p} - J \tag{2.29}$$

$$\text{Annual inventory holding cost} = \frac{V}{2}(t_2 + t_3)\frac{R}{yQ}\cdot PF$$

As
$$t_2 + t_3 = \frac{V}{(yp-r)} + \frac{V}{r} = \frac{ypV}{r(yp-r)} periods = \frac{ypV}{r(yp-r)}\cdot\frac{r}{R} year = \frac{ypV}{R(yp-r)} year,$$

$$\text{Annual holding cost} = \frac{ypV^2}{2R(yp-r)}\cdot\frac{R}{yQ}\cdot PF = \frac{pV^2PF}{2Q(yp-r)}$$

Substituting the value of V from Equation 2.29, annual inventory holding cost:

$$= \frac{pPF}{2Q(yp-r)}\left[\frac{(yp-r)Q}{p} - J\right]^2$$

$$= \frac{(yp-r)PFQ}{2p} - PFJ + \frac{PFpJ^2}{2(yp-r)Q} \tag{2.30}$$

$$\text{Annual production and setup cost} = \frac{R}{y}\cdot P + \frac{R}{yQ}\cdot C \tag{2.31}$$

Adding Equation 2.28, Equation 2.30) and Equation 2.31, total relevant cost:

$$E = \frac{pJ^2(K+PF)}{2(yp-r)Q} + \frac{(yp-r)PFQ}{2p} - PFJ + \frac{RP}{y} + \frac{RC}{yQ} \tag{2.32}$$

$$\text{Now, } \frac{\partial E}{\partial J} = \frac{pJ(K+PF)}{(yp-r)Q} - PF = 0$$

$$\text{Or } J = \frac{PF(yp-r)Q}{p(K+PF)} \tag{2.33}$$

$$\text{And } \frac{\partial E}{\partial Q} = -\frac{pJ^2(K+PF)}{2(yp-r)Q^2} + \frac{(yp-r)PF}{2p} - \frac{RC}{yQ^2} = 0$$

$$\text{Or } Q^2 = \frac{yp^2J^2(K+PF) + 2RCp(yp-r)}{yPF(yp-r)^2}$$

Substituting the value of J from Equation 2.33,

$$Q^* = \sqrt{\frac{2RCp(K+PF)}{(yp-r)yPFK}} \tag{2.34}$$

Substituting optimal Q from Equation 2.34 in Equation 2.33,

$$J^* = \sqrt{\frac{2RCPF(yp-r)}{ypK(K+PF)}} \tag{2.35}$$

Optimal total cost is obtained by substituting Equation 2.34 and Equation 2.35 in Equation 2.32. As discussed in Section 2.1.1, conditions for optimality given by Equation 2.10 and Equation 2.11 will need to be satisfied.

Now, $\frac{\partial^2 E}{\partial J^2} = \frac{p(K+PF)}{(yp-r)Q}$ which is > 0, and therefore, Condition 2.11 is satisfied.

$$\frac{\partial^2 E}{\partial Q^2} = \frac{pJ^2(K+PF)}{(yp-r)Q^3} + \frac{2RC}{yQ^3}$$

and $\frac{\partial^2 E}{\partial Q \partial J} = \frac{-pJ(K+PF)}{(yp-r)Q^2}$

From Condition 2.10, $\frac{\partial^2 E}{\partial Q^2} \cdot \frac{\partial^2 E}{\partial J^2} > \left[\frac{\partial^2 E}{\partial Q \partial J}\right]^2$

Substituting the values,

$$\frac{p(K+PF)}{(yp-r)Q}\left[\frac{pJ^2(K+PF)}{(yp-r)Q^3} + \frac{2RC}{yQ^3}\right] > \frac{p^2J^2(K+PF)^2}{(yp-r)^2Q^4}$$

Or $\frac{2RCp(K+PF)}{y(yp-r)Q^4} > 0$, which is true, and therefore the solutions obtained are optimal.

Maximum inventory during the cycle, V is an important parameter to know, so that storage capacity for the items can be ensured or intermediate storage space between two machines can be managed. Optimal maximum inventory is evaluated by substituting Q* and J* from Equation 2.34 and Equation 2.35 respectively, into Equation 2.29.

Example 2.5

Consider the data for Example 2.4 which are: R = 12,000, C = Rs. 200, y = 0.98, F = 0.2, P = Rs. 10, r = 1,000, p= 1,200

Assume additional parameter, i.e. annual shortage cost per unit K as Rs. 400. Obtain the optimal production lot size, optimal shortage quantity and total relevant cost. Compare the results with that of Example 2.4 and comment.

From Equation 2.34, $Q^* = \sqrt{\frac{2RCp(K+PF)}{(yp-r)yPFK}}$

$$= \sqrt{\frac{2 \times 12000 \times 200 \times 1200(400+2)}{(0.98 \times 1200 - 1000) \times 0.98 \times 10 \times 0.2 \times 400}}$$

= 4,096.47 units

From Equation 2.35, $J^* = \sqrt{\frac{2RCPF(yp-r)}{ypK(K+PF)}}$

$$= \sqrt{\frac{2 \times 12000 \times 200 \times 10 \times 0.2(1176-1000)}{0.98 \times 1200 \times 400(400+2)}}$$

= 2.99 units

From Equation 2.32, total relevant cost,

$$E = \frac{pJ^2(K+PF)}{2(yp-r)Q} + \frac{(yp-r)PFQ}{2p} - PFJ + \frac{RP}{y} + \frac{RC}{yQ}$$

Substituting Q* and J* from Equation 2.34 and 2.35 Equation, and on solving,

$$E^* = PF\sqrt{\frac{RCPF(yp-r)}{2ypK(K+PF)}} + \sqrt{\frac{RCPF(yp-r)(K+PF)}{2ypK}} - PF\sqrt{\frac{2RCPF(yp-r)}{ypK(K+PF)}} + \sqrt{\frac{RCPFK(yp-r)}{2yp(K+PF)}} + \frac{RP}{y}$$

Or

$$E^* = \sqrt{\frac{RCPF(yp-r)(K+PF)}{2ypK}} - PF\sqrt{\frac{RCPF(yp-r)}{2ypK(K+PF)}} + \sqrt{\frac{RCPFK(yp-r)}{2yp(K+PF)}} + \frac{RP}{y}$$

Or $E^* = \sqrt{\frac{KRCPF(yp-r)}{2yp(K+PF)}} + \sqrt{\frac{RCPFK(yp-r)}{2yp(K+PF)}} + \frac{RP}{y}$

Or $E^* = \sqrt{\frac{2RCPFK(yp-r)}{yp(K+PF)}} + \frac{RP}{y}$ (2.36)

Substituting the numerical values, optimum total annual cost, E* = Rs. 123,644.63.

Total cost in Example 2.4 was obtained as Rs. 123,647.62 when shortages were not considered. Therefore, a cost reduction is achieved with increase in optimal production lot size.

When shortages are not allowed, optimum total cost is obtained by substituting Equation 2.27 in Equation 2.26. Call this as E_1* and

$$E_1^* = \sqrt{\frac{2RCPF(yp-r)}{yp}} + \frac{RP}{y} \tag{2.37}$$

Subtracting Equation 2.36 from Equation 2.37:

$$E_1^* - E^* = \sqrt{\frac{2RCPF(yp-r)}{yp}}\left[1-\sqrt{\frac{K}{(K+PF)}}\right] \tag{2.38}$$

As K/(K + PF) is less than 1, E_1*–E* is positive and cost saving is obtained by allowing shortages in the production system. However, as discussed in Example 2.2 of Section 2.1.1, a decision to allow shortages also depends on appropriate service level.

As the shortage cost was assumed to be very high as compared to unit production cost, optimal shortage quantity J* evaluated is very less. Obviously, cost savings are also less.

2.2.2 Fractional Backordering

As discussed in Section 2.1.2, a common parameter concerning shortage cost is associated with shortage quantities, whether these result into backordering or not. Now, with reference to Figure 2.4 and as explained in Section 2.2.1,

$$\text{Annual shortage quantity} = \frac{pJ^2}{2(yp-r)Q}$$

Since a fraction b of shortage quantity is not backordered,

$$\text{Annual production cost} = \left(\frac{R}{y} - \frac{bpJ^2}{2(yp-r)Q}\right)P \tag{2.39}$$

$$\text{Yearly setup cost} = \left(\frac{R}{y} - \frac{bpJ^2}{2(yp-r)Q}\right)\frac{C}{Q} \tag{2.40}$$

$$\text{Annual shortage cost} = \frac{pJ^2K}{2(yp-r)Q} \tag{2.41}$$

As derived in Section 2.2.1,

$$\text{Annul holding cost} = \frac{(yp-r)PFQ}{2p} - PFJ + \frac{PFpJ^2}{2(yp-r)Q} \tag{2.42}$$

Adding Equations 2.39–2.42, total relevant cost:

$$E = \frac{pJ^2(K+PF)}{2(yp-r)Q} + \frac{(yp-r)PFQ}{2p} - PFJ + \left[\frac{R}{y} - \frac{bpJ^2}{2(yp-r)Q}\right]$$
$$P + \frac{RC}{yQ} - \frac{bpC}{2(yp-r)}\left[\frac{J}{Q}\right]^2 \tag{2.43}$$

The assumption of invariance of frequency of ordering was not considered a serious assumption, as discussed before. Similarly, the last term of Equation 2.43, which relates to the change in the frequency of setups due to fractional backordering, may be ignored. Therefore, an almost optimal solution is obtained for all practical purposes using the following equation for total cost:

$$E = \frac{(K+PF-bP)pJ^2}{2(yp-r)Q} + \frac{RP}{y} + \frac{(yp-r)PFQ}{2p} - PFJ + \frac{RC}{yQ} \tag{2.44}$$

Now, $\frac{\partial E}{\partial J} = 0$ shows

$$J = \frac{PF(yp-r)Q}{p(K+PF-bP)} \tag{2.45}$$

And $\frac{\partial E}{\partial Q} = 0$ shows

$$\frac{(K+PF-bP)pJ^2}{2(yp-r)Q^2} + \frac{RC}{yQ^2} = \frac{(yp-r)PF}{2p}$$

Substituting the value of J from Equation 2.45,

$$\frac{P^2F^2(yp-r)}{2p(K+PF-bP)} + \frac{RC}{yQ^2} = \frac{(yp-r)PF}{2p}$$

$$\text{Or } \frac{RC}{yQ^2} = \frac{PF(yp-r)(K-bP)}{2p(K+PF-bP)}$$

$$\text{Or } Q^* = \sqrt{\frac{2pRC(K+PF-bP)}{PFy(yp-r)(K-bP)}} \tag{2.46}$$

Substituting Equation 2.46 in Equation 2.45:

$$J^* = \sqrt{\frac{2RCPF(yp-r)}{yp(K+PF-bP)(K-bP)}} \tag{2.47}$$

To ensure the optimality,

$$\frac{\partial^2 E}{\partial J^2} = \frac{p(K + PF - bP)}{(yp - r)Q}$$

which is greater than zero, as K > P in the real environment, and Condition 2.11 is satisfied.

From Condition 2.10, $\frac{\partial^2 E}{\partial Q^2} \cdot \frac{\partial^2 E}{\partial J^2} > \left[\frac{\partial^2 E}{\partial Q \partial J}\right]^2$

Or
$$\left[\frac{(K + PF - bP)pJ^2}{(yp - r)Q^3} + \frac{2RC}{yQ^3}\right]\frac{p(K + PF - bP)}{(yp - r)Q} > \left[\frac{-pJ(K + PF - bP)}{(yp - r)Q^2}\right]^2$$

Or $\frac{2RCp(K + PF - bP)}{y(yp - r)Q^4} > 0$, which is true as K > P and because b—as well as F—are fractions. Since both the conditions are satisfied with K > P, Equation 2.46 and Equation 2.47 give optimal solutions. Substituting these equations in Equation 2.44, optimum total annual cost,

$$E^* = PF\sqrt{\frac{RCPF(yp - r)}{2yp(K - bP)(K + PF - bP)}} + \frac{RP}{y} + \sqrt{\frac{RCPF(yp - r)(K + PF - bP)}{2yp(K - bP)}} - PF\sqrt{\frac{2RCPF(yp - r)}{yp(K + PF - bP)(K - bP)}} + \sqrt{\frac{RCPF(yp - r)(K - bP)}{2yp(K + PF - bP)}}$$

Or
$$E^* = \frac{\left[PF + (K + PF - bP) - 2PF + (K - bP)\right]\sqrt{RCPF(yp - r)}}{\sqrt{2yp(K - bP)(K + PF - bP)}} + \frac{RP}{y}$$

Or $$E^* = \frac{2(K - bP)\sqrt{RCPF(yp - r)}}{\sqrt{2yp(K - bP)(K + PF - bP)}} + \frac{RP}{y}$$

Or $$E^* = \sqrt{\frac{2RCPF(yp - r)(K - bP)}{yp(K + PF - bP)}} + \frac{RP}{y} \tag{2.48}$$

Example 2.6

Consider the following data:

R = 12,000, C = Rs. 200, y = 0.98, F = 0.2, P = Rs. 10, r = 1,000, p = 1,200, K = Rs. 100 and b = 0.3

Evaluate the optimum values of production lot size, maximum shortage quantity and total annual cost. Also, show how different these values are from those obtained when change in frequency of setups due to fractional backordering is taken into consideration. Comment on the results.

From Equation 2.46, $Q^* = \sqrt{\dfrac{2\times1200\times12000\times200(100+2-3)}{10\times0.2\times0.98(0.98\times1200-1000)(100-3)}}$

$= 4{,}128.18$ units

From Equation 2.47, $J^* = \sqrt{\dfrac{2\times12000\times200\times10\times0.2(0.98\times1200-1000)}{0.98\times1200(100+2-3)(100-3)}}$

$= 12.23$ units

From Equation 2.48, E* = Rs. 123,635.45

When change in frequency of setups due to fractional backordering is taken into consideration, total annual cost is given by Equation 2.43, i.e.,

$$E = \frac{pJ^2(K+PF)}{2(yp-r)Q} + \frac{(yp-r)PFQ}{2p} - PFJ + \left[\frac{R}{y} - \frac{bpJ^2}{2(yp-r)Q}\right] P + \frac{RC}{yQ} - \frac{bpCJ^2}{2(yp-r)Q^2}$$

Substituting numerical values,

$$E(Q,J) = \frac{3825J^2}{11Q} + \frac{11Q}{75} - 2J + 122448.98 - \frac{225J^2}{22Q} + \frac{12X10^7}{49Q} - \frac{2250J^2}{11Q^2}$$

Or

$$E(Q,J) = 122448.98 + \frac{7425J^2}{22Q} + \frac{12X10^7}{49Q} + \frac{11Q}{75} - \frac{2250J^2}{11Q^2} - 2J \tag{2.49}$$

Applying the univariate method in which each variable in turn is considered and only one variable is changed at a time in order to search for optimum. Substituting Q = 4,128.18 and J = 12.23 in Equation 2.49, E (4,128.18, 12.23)

= Rs. 123,635.45.

To observe the effect of change in J value, consider only those components of Equation 2.49 which include J,

$$\text{Say } \varphi(Q,J) = \frac{7425J^2}{22Q} - \frac{2250J^2}{11Q^2} - 2J$$

If ϕ (Q, J) is less, it should be preferred because it contributes to reduction in the cost.

Now ϕ (4,128.18, 12.23) = −12.233439

Changing the J by ± 0.01 from the existing value,

ϕ (4,128.18, 12.22) = −12.233426 > −12.233439

And ϕ (4,128.18, 12.24) = −12.233437 > −12.233439

As the change in both the directions tend to increase the cost further, existing value of J is preferred. Now keeping the J = 12.23 as fixed, vary Q in both the positive and negative directions by 0.1

$$\varphi(Q,J) = \frac{7425J^2}{22Q} + \frac{12X10^7}{49Q} + \frac{11Q}{75} - \frac{2250J^2}{11Q^2} - 2J$$

ϕ (4,128.18, 12.23) = 1,186.4677

ϕ (4,128.28, 12.23) = 1,186.482

and ϕ (4,128.08, 12.23) = 1,186.4677,

Similarly decreasing the value of Q further, it may be observed that no appreciable improvement in the cost occurs, and for all practical purposes:

Optimal, Q = 4,128.18 units

J = 12.23 units

E = Rs. 123,635.45,

which were the solutions obtained by Equations 2.46–2.48.

Observing Equation 2.49, the additional feature due to consideration of change in frequency of setups is $\frac{-2250J^2}{11Q^2} = -1.795 \times 10^{-3}$, which is negligible—and therefore, the solutions do not differ practically.

Exercises

1. If the total cost concerning purchase of a component is very high, what should be the approach of a company to solve this problem?
2. Derive the economic procurement lot size with inclusion of quality defects considering the following case:

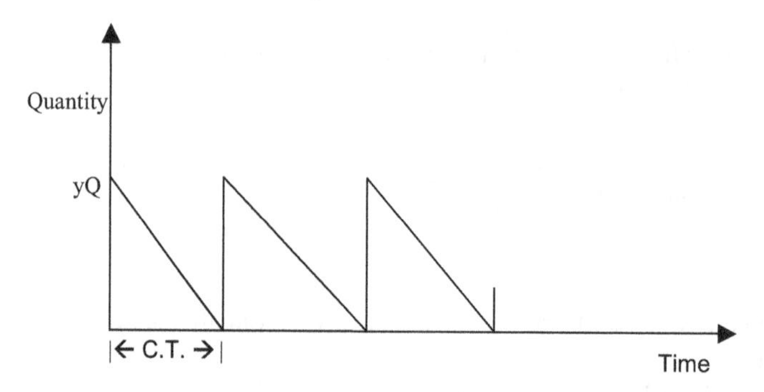

3. Consider the following data, and determine the economic procurement lot size, optimum total annual cost and cycle time in periods:

   Annual demand for nondefective items = 24,000
   Demand rate per period (say per month) = 2,000
   Fixed ordering cost per order = Rs. 100
   Unit purchase cost = Rs. 20
   Annual inventory carrying cost fraction = 0.25
   Proportion of nondefective items = 0.98

4. What do you understand by the shortages?
5. With the inclusion of shortages and quality defects, derive for the optimum values of the following parameters:

   a. Procurement lot size
   b. Maximum shortage quantity

6. Assume an additional parameter, annual shortage cost per unit as Rs. 50, along with the following data; obtain optimal ordering and shortage quantity, and also evaluate the optimum total annual cost:

   Annual demand for nondefective items = 24,000
   Proportion of nondefective items = 0.96
   Fixed ordering cost per order = Rs. 120
   Unit purchase cost = Rs. 15
   Annual inventory carrying cost fraction = 0.25

7. In the previous exercise, evaluate the optimum total annual cost if shortages are not included. Compare both the cases and comment. Also derive the difference in cost in general for purchase situation.

8. Consider the following scenario:

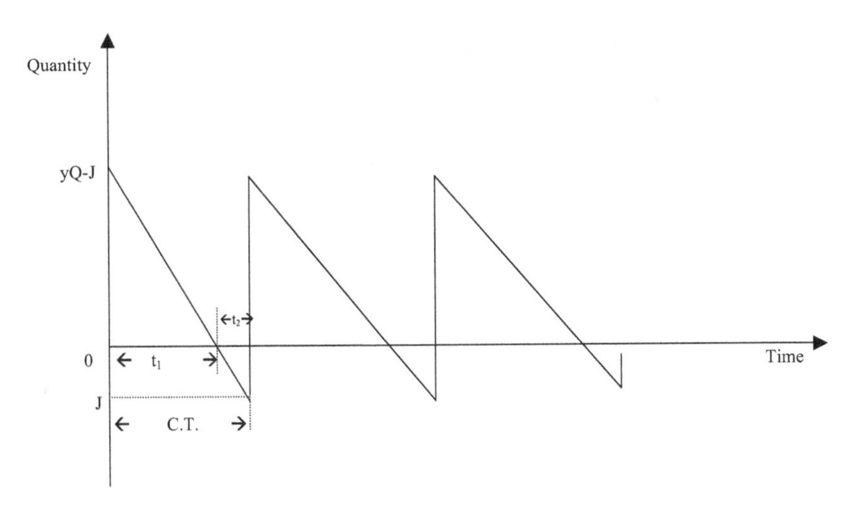

If partial backordering is additionally included, then derive the following annual costs after explaining the partial backordering:

a. Shortage cost
b. Purchase cost
c. Ordering cost
d. Inventory holding cost

Express total relevant cost and optimize.

9. Input information is given below:

Annual shortage cost per unit = Rs. 60
Annual demand for nondefective items = 24,000
Proportion of nondefective items in a batch = 0.97
Fixed ordering cost per order = Rs. 100
Unit purchase cost = Rs. 20
Annual inventory carrying cost fraction = 0.3
Fraction of shortage quantity which is not backordered = 0.2

Obtain optimal ordering and shortage quantity. Also evaluate the optimum total annual cost.

10. What do you understand by the batch production?
    Consider the following situation; after explaining this, derive the economic production quantity with inclusion of quality defects:

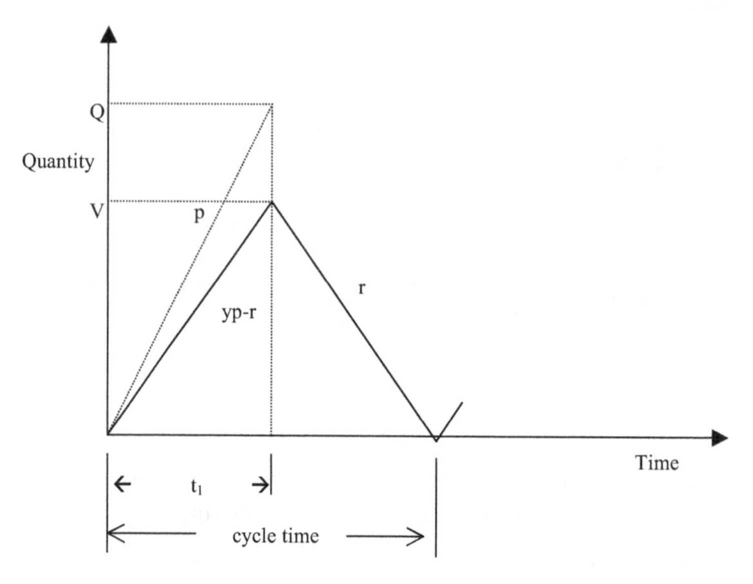

11. Evaluate: (a) economic production or batch quantity, and (b) cycle time in periods for the input parameters as provided below:

    Production cost per unit = Rs. 15
    Demand rate per period = 1,000 units
    Production rate per period = 1,250 units
    Annual demand = 12,000 units
    Setup cost = Rs. 250
    Proportion of nondefective items in a lot = 0.99
    Annual inventory carrying cost fraction = 0.25

    Also, conduct the sensitivity analysis of optimum values of lot size and total cost with respect to a scenario when proportion of nondefective items in a lot is equal to 1, i.e. when quality defects are not incorporated.

12. Consider the following scenario:

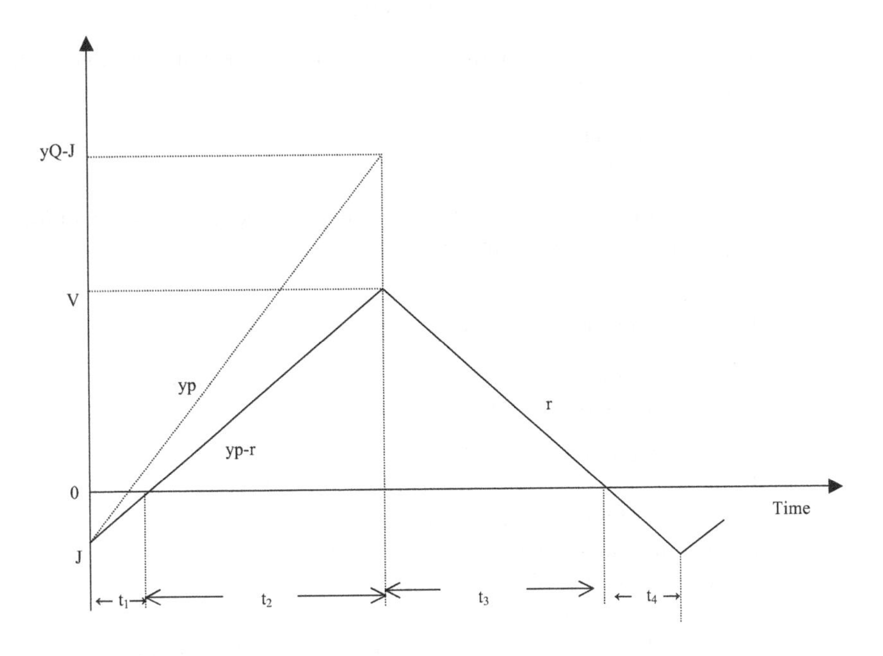

Derive the following annual costs pertaining to:

a. Shortage
b. Inventory holding
c. Production
d. Setup

After adding all the cost components just listed, express the total cost and optimize to get the following output parameters:

a. Batch size
b. Maximum shortage quantity

Also, obtain the optimal total cost expression in generalized form. Compare with the situation when shortages are not allowed and comment about the cost savings.

13. Evaluate: (a) economic production or batch quantity, and (b) maximum shortage quantity for the input parameters as follows:

Production cost per unit = Rs. 12
Demand rate per period = 1,000 units
Production rate per period = 1,250 units
Annual shortage cost = Rs. 350
Annual demand = 12,000 units
Setup cost = Rs. 250

Proportion of nondefective items in a lot = 0.97
Annual inventory carrying cost fraction = 0.3

14. Consider Exercise 12. Additionally include the fractional backordering in the production-inventory system and show the effect on:

    a. Yearly production cost
    b. Yearly setup cost

    After adding all the cost components, express the total cost and optimize to get the following output parameters:

    a Batch size
    b. Maximum shortage quantity

    Also, obtain the optimal total cost expression in a generalized form.
15. Assume the following data:

    Production cost per unit = Rs. 15
    Demand rate per period = 1,000 units
    Proportion of nondefective items in a lot = 0.99
    Annual inventory carrying cost fraction = 0.2
    Production rate per period = 1,300 units
    Annual shortage cost = Rs. 400
    Fraction of shortage quantity which is not backordered = 0.2
    Annual demand = 12,000 units
    Setup cost = Rs. 300

    Using these data, evaluate the optimal output parameters such as:

    a. Production batch size
    b. Maximum shortage quantity
    c. Total annual cost

3 Batch Size Relevance for Just-in-Time (JIT)

In the just-in-time (JIT) philosophy, the emphasis is on keeping inventories as small as possible. Consider a company engaged in manufacturing of a product which requires a certain input item. The management is interested in procuring the minimum stock of the input item. A model needs to be developed for this purpose, including production of finished item and replenishment of the input item.

3.1 PRODUCTION WITH REPLENISHMENT OF INPUT ITEM

For a successful practical implementation of the JIT concept, there has to be a strategic alliance between the manufacturing organization and the supplier of the input item. Therefore, it is assumed that rigorous inspection procedures are adopted at the supplier's end and no defective input item would be supplied to the manufacturing company. Production of end item, along with replenishment of input item, is shown in Figure 3.1.

Let:

Annual demand for end item = R
Demand rate of end item per period = r
Production rate/period = p
Proportion of nondefective end items in a lot = y
Production lot size = Q
Units of the input item required per unit of the end product = n
Replenishmentrate of the input item = d
Number of orders for procurement of the input item in a production cycle = k
Fixed ordering cost for the input item = A
Fixed production setup cost for the end product = C
Production cost per unit (including value addition) = P
Annual inventory carrying cost fraction = F
Purchase cost per unit of the input item = G

Total annual cost concerning the end product was obtained in Section 2.2, as expressed by Equation 2.26, i.e.,

$$= \frac{(yp-r)PFQ}{2p} + \frac{RC}{yQ} + \frac{PP}{y} \tag{3.1}$$

Referring to Figure 3.1, cycle time is yQ/r and production takes place for time Q/p in the C.T. Input item procurement is during time Q/p only.

DOI: 10.1201/9781003213994-3

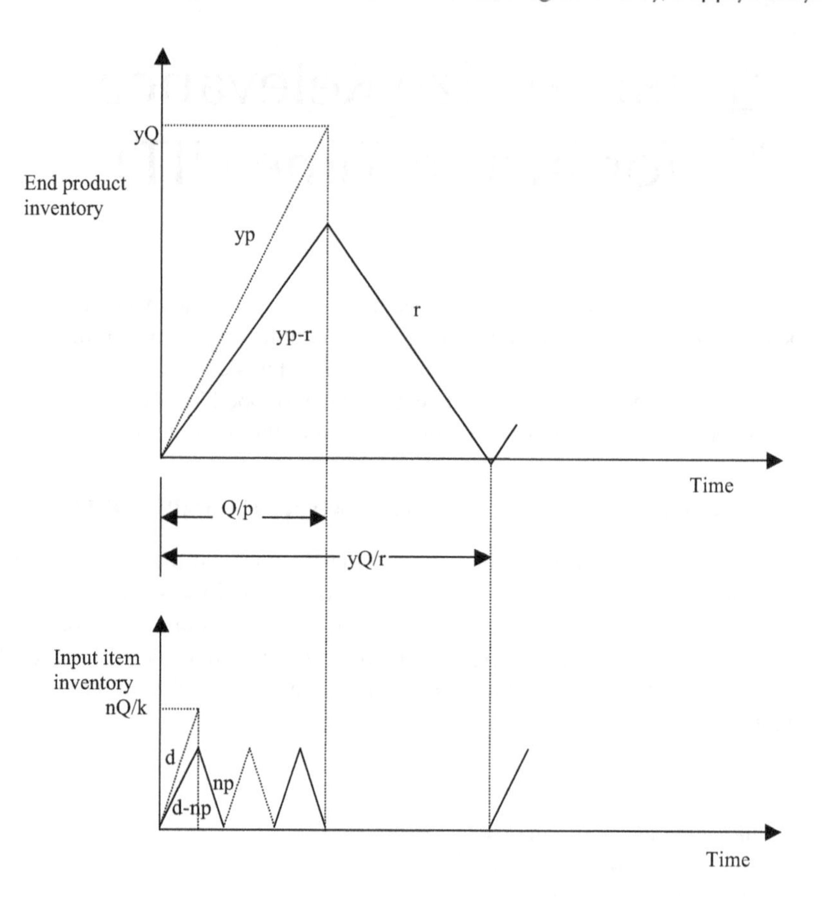

FIGURE 3.1 Production of end item, along with replenishment of the input item.

$$\text{Fraction of cycle time during which input item inventory exists} = \frac{(Q/p)}{(yQ/r)} = \frac{r}{yp}$$

As Q is the production lot size and n units of input inventory item are required per unit of the end product, requirement of the input item during the cycle is nQ (for example, if 3 kg of the input item are required for one end item, then n = 3). As frequent replenishment of input item is there, which is consistent with JIT concept, and k is the frequency of ordering, procurement lot size is (nQ/k).

Since production rate is p, consumption rate of the input item is np. Rate of replenishment is d, which is greater than np, and (d – np) is the inventory build-up rate. 'Zero inventory' is an ideal case in which d = np.

$$\text{Maximum input item inventory} = \frac{(d-np)(nQ/k)}{d}$$

and average inventory $= \dfrac{(d-np)nQ}{2kd}$

which exists for a fraction of cycle time, i.e. r/yp.

Annual input item inventory holding cost $= \dfrac{(d-np)nQ}{2kd} \cdot \dfrac{r}{yp} \cdot GF$

where G is the purchase cost per unit of the input item and F is the annual holding cost fraction.

As there are (R/yQ) production cycles in one year, the annual number of orders for input item is (Rk/yQ).

Annual input item ordering cost $= \dfrac{Rk}{yQ} \cdot A$

Annual cost concerning the input item $= \dfrac{RkA}{yQ} + \dfrac{(d-np)nQrGF}{2kdyp}$ (3.2)

Adding Equation 3.1 and Equation 3.2, total relevant cost:

$$E = \frac{(yp-r)PFQ}{2p} + \frac{RP}{y} + (C+kA)\frac{R}{yQ} + \frac{(d-np)nQrGF}{2kdyp} \quad (3.3)$$

$$\partial E/\partial k = 0, \text{ shows } k = Q\sqrt{\frac{(d-np)nrGF}{2dpAR}} \quad (3.4)$$

$$\partial E/\partial Q = 0 \text{ shows, } \frac{CR}{yQ^2} + \frac{kAR}{yQ^2} = \frac{(yp-r)PF}{2p} + \frac{(d-np)nrGF}{2kdyp}$$

Substituting the value of k from Equation 3.4 and on solving,

$$\text{Optimum } Q^* = \sqrt{\frac{2pCR}{(yp-r)PFy}} \quad (3.5)$$

Substituting Q* in Equation 3.4,

$$k^* = \sqrt{\frac{(d-np)nrGC}{(yp-r)PydA}} \quad (3.6)$$

Optimum total cost is obtained by substituting Equation 3.5 and Equation 3.6 in Equation 3.3.

To provide more generalism, replenishment rate of the input item is considered to be finite. This may be useful in some cases. For instance, some of the steel tube manufacturers have set up their own cold rolling mills in order to obtain the cold rolled steel sheets in desired specifications as an input item for further processing. The present analysis is helpful in synchronizing the replenishment rate with the production of finished item along with the overall objective of total cost optimization. However in many cases, procurement of the input item in smaller lot sizes is instantaneous, as discussed in the next section.

3.2 INSTANTANEOUS PROCUREMENT

This situation is shown in Figure 3.2. Frequent ordering of the input item during production time takes place. Replenishment rate d is infinite. In other words, procurement of raw material/input item is instantaneous. Equation 3.3 can be adjusted for this situation as follows:

$$E = \frac{(yp - r)PFQ}{2p} + \frac{RP}{y} + (C + kA)\frac{R}{yQ} + \frac{nQrGF}{2kyp} \tag{3.7}$$

Following the similar procedure, optimal values of production lot size and ordering frequency are obtained as follows:

$$Q^* = \sqrt{\frac{2pCR}{(yp - r)PFy}} \tag{3.8}$$

$$\text{and } k^* = \sqrt{\frac{nrGC}{(yp - r)PyA}} \tag{3.9}$$

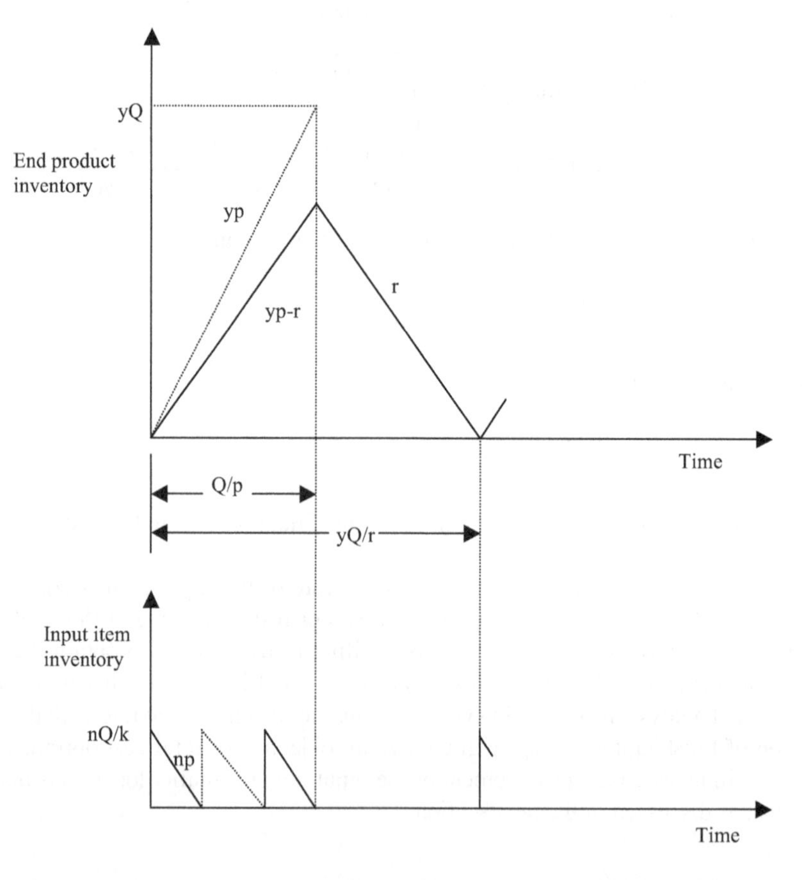

FIGURE 3.2 Instantaneous procurement of the input item.

The conditions for optimality regarding the two-variables optimization problem can easily be satisfied for this case, as well as for the previous situation in which the replenishment rate was finite.

The optimal total cost, E*, is obtained by substituting Equation 3.8 and Equation 3.9 in Equation 3.7.

$$E^* = \frac{RP}{y} + \sqrt{\frac{2(yp-r)PFCR}{yp}} + \frac{1}{y}\sqrt{\frac{2nrGFAR}{p}} \tag{3.10}$$

In the preceding equation, P includes the value addition. For instance, if manufacturing cost alone per unit is P', then P = P'+ nG.

Example 3.1

Consider the following parameters:

Annual demand, R = 1,200
Production setup cost, C = Rs. 200
Annual inventory holding cost fraction, F = 0.2
Demand rate of end item per period, r = 100
Production rate per period, p = 120
Proportion of acceptable end items in a lot, y = 0.95
Unit purchase cost of the input item, G = Rs. 10
Units of the input item required per unit of end item, n = 2
Unit production cost (including value addition), P = Rs. 30
Fixed ordering cost for the input item, A = Rs. 50

Obtain the optimum values of production lot size, frequency of ordering of the input item and total cost.

Now, from Equation 3.8, $Q^* = \sqrt{\frac{2\times120\times200\times1200}{(0.95\times120-100)\times30\times0.2\times0.95}}$

$= 849.59$ units

From Equation 3.9, $k^* = \sqrt{\frac{2\times100\times10\times200}{(114-100)\times30\times0.95\times50}}$

$= 4.48$

and from Equation 3.10, E* = Rs. 39,155.19

For the implementation of the preceding results, the value of k* will need to be converted to the nearest integer. However, for the sensitivity analysis or for determining the effects of various input parameters, fractional values of k* may be used.

3.2.1 Sensitivity Analysis

In the JIT environment, the emphasis is on reduction in ordering costs. Similarly, in order to produce in small lot sizes, setup cost needs to be reduced. It is of interest to conduct the sensitivity analysis corresponding to ordering and setup cost.

Consider the data for Example 3.1, ordering cost A (Rs. 50) is decreased by 10%, 20% and 30% from its present value, i.e. corresponding values of A are Rs. 45, Rs. 40 and Rs. 35, respectively. For each value of A, total optimal cost, E_1* is obtained using Equation 3.10. Percentage decrease in the cost is obtained by computing (E* – E_1*) × 100/E* where E* is Rs. 39,155.19 as evaluated in Example 3.1.

A (Rs.)	% Decrease in A	E1* (Rs.)	% Decrease in Cost [(E*–E_1*) × 100/E*]
45	10%	39,121.03	0.09%
40	20%	39,084.91	0.18%
35	30%	39,046.45	0.28%

Following the similar procedure, effects of decrease in setup cost can be observed. At present, the setup cost C is Rs. 200. By improvement in the technology and current practices, value of C may be decreased. The following table shows these effects. Percentage decrease in cost with respect to 10%, 20% and 30% decrease in C are 0.08%, 0.16% and 0.25%, respectively.

C (Rs.)	% Decrease in C	E_1* (Rs.)	% Decrease in Cost [(E*–E_1*) × 100/E*]
180	10%	39,124.68	0.08%
160	20%	39,092.41	0.16%
140	30%	39,058.05	0.25%

Although percentage decrease in cost will depend on the values of input parameters, this decrease is, generally speaking, more sensitive for higher decrease in ordering and setup cost. Effects on the cost are more related to decrease in ordering cost as compared to decrease in setup cost. Given the choice, efforts should be made to decrease the ordering cost first, and then setup cost reduction should be undertaken.

3.3 INCLUDING SHORTAGES

As discussed in Chapter 2, shortages may be allowed in the system. Shortage quantity is assumed to be completely backordered. Figure 3.3 shows this case. During the production time Q/p in the cycle, procurement of the input item in frequent lot sizes takes place. No procurement exists during the down time in the production-inventory cycle.

Total cost expression concerning end item was developed in Section 2.2.1 given by Equation 2.32; this is stated in what follows using S as shortage cost instead of K:

$$\text{Cost concerning end item} = \frac{pJ^2(S+PF)}{2(yp-r)Q} + \frac{(yp-r)PFQ}{2p} - PFJ + \frac{RP}{y} + \frac{RC}{yQ} \tag{3.11}$$

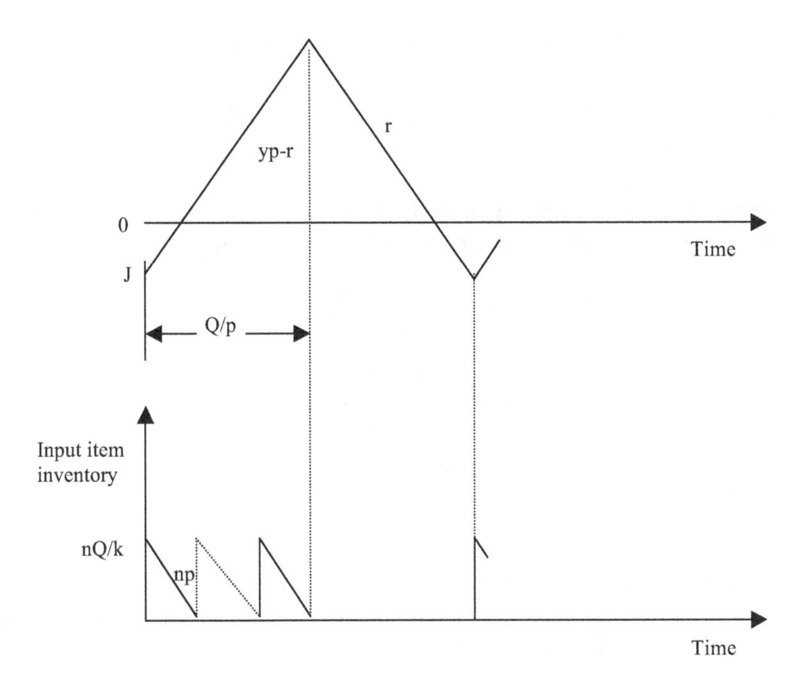

FIGURE 3.3 Including shortages in the production with procurement of the input item.

Where:

J = maximum shortage quantity
S = shortage cost per unit-year
P includes value addition, as explained before

Cost concerning the input item was expressed by Equation 3.2, i.e.

$$\frac{RkA}{yQ} + \frac{(d - np)nQrGF}{2kdyp}$$

which can be written as

$$\frac{RkA}{yQ} + \frac{(1 - np / d)nQrGF}{2kyp}$$

For instantaneous procurement, d = ∞ and therefore (1 − np/d) = 1 and annual cost concerning the input item:

$$= \frac{RkA}{yQ} + \frac{nQrGF}{2kyp} \tag{3.12}$$

Adding Equation 3.11 and Equation 3.12, total relevant cost:

$$E = \frac{pJ^2(S+PF)}{2(yp-r)Q} + \frac{(yp-r)PFQ}{2p} - PFJ + \frac{RP}{y} + \frac{RC}{yQ} + \frac{RkA}{yQ} + \frac{nQrGF}{2kyp} \tag{3.13}$$

Differentiating the preceding equation partially with respect to J, k and Q, and equating to zero, following optimal values are obtained on solving:

$$Q^* = \sqrt{\frac{2RCp(S+PF)}{SPFy(yp-r)}} \tag{3.14}$$

$$J^* = \sqrt{\frac{2RCPF(yp-r)}{Spy(S+PF)}} \tag{3.15}$$

and $$k^* = \sqrt{\frac{(S+PF)nrGC}{(yp-r)PyAS}} \tag{3.16}$$

By substituting values from Equations 3.14–3.16 in Equation 3.13, total cost E* can be obtained. As it is a three variables optimization (minimization) problem, the following conditions will need to be satisfied for optimality.

$$A > 0 \tag{3.17}$$

$$AB > H^2 \tag{3.18}$$

$$ABC + 2FGH - AF^2 - BG^2 - CH^2 > 0 \tag{3.19}$$

Where $A = \frac{\partial^2 E}{\partial Q^2}, B = \frac{\partial^2 E}{\partial J^2}, C = \frac{\partial^2 E}{\partial k^2}, H = \frac{\partial^2 E}{\partial Q \partial J}, F = \frac{\partial^2 E}{\partial J \partial k}, G = \frac{\partial^2 E}{\partial k \partial Q}$

Now $A = \frac{\partial^2 E}{\partial Q^2} = \frac{pJ^2(S+PF)}{(yp-r)Q^3} + \frac{2(C+kA)R}{yQ^3}$

which is positive and therefore Condition 3.17 is satisfied.

$$B = \frac{\partial^2 E}{\partial J^2} = \frac{p(S+PF)}{(yp-r)Q}$$

$$H = \frac{\partial^2 E}{\partial Q \partial J} = \frac{-pJ(S+PF)}{(yp-r)Q^2}$$

Second Condition 3.18 is, $AB > H^2$

Or $$\left[\frac{pJ^2(S+PF)}{(yp-r)Q^3} + \frac{2(C+kA)R}{yQ^3}\right]\frac{p(S+PF)}{(yp-r)Q} > \frac{p^2J^2(S+PF)^2}{(yp-r)^2Q^4}$$

$$\text{Or } \frac{2(C+kA)Rp(S+PF)}{y(yp-r)Q^4} > 0$$

Which is true and therefore this condition is also satisfied.

$$\text{Now, } C = \frac{\partial^2 E}{\partial k^2} = \frac{nQrGF}{ypk^3}$$

$$G = \frac{\partial^2 E}{\partial k \partial Q} = -\frac{RA}{yQ^2} - \frac{nrGF}{2ypk^2}$$

As $F = \frac{\partial^2 E}{\partial J \partial k} = 0$, the Condition 3.19 reduces to ABC – BG² – CH² > 0,

Substituting the values and on solving,

$$\frac{2RnrGF(C+kA)}{y^2 pQ^2k^3} > \left[\frac{RA}{yQ^2} + \frac{nrGF}{2ypk^2}\right]^2 \tag{3.20}$$

From Equation 3.14 and Equation 3.16,

$$k = Q\sqrt{\frac{nrGF}{2RAp}},$$

Substituting this value of k in Equation 3.20,

$$\frac{4R^2CA}{y^2Q^5}\sqrt{\frac{2RAp}{nrGF}} + \left[\frac{2RA}{yQ^2}\right]^2 > \left[\frac{2RA}{yQ^2}\right]^2$$

$$\text{Or } \frac{4R^2CA}{y^2Q^5}\sqrt{\frac{2RAp}{nrGF}} > 0$$

As the left-hand side of the equation (L.H.S.) is positive, Condition 3.19) is satisfied and therefore Equations 3.14–3.16 represent optimal values. Substituting these optimal results in Equation 3.13 and on solving,

$$E^* = \frac{RP}{y} + \frac{1}{y}\sqrt{\frac{2RAnrGF}{p}} + \sqrt{\frac{2PFRCS(yp-r)}{yp(S+PF)}} \tag{3.21}$$

Example 3.2

Consider the data for Example 3.1, i.e.,

R = 1,200, C = Rs. 200, F = 0.2, R = 100, p = 120, y = 0.95, G = Rs. 10, n = 2, P = Rs. 30, A = Rs. 50

Assume additional value for annual shortage cost, S = Rs. 200. Compute the optimal values of production quantity, shortage quantity, number of cycles for procurement of the input item and annual cost. Comment on the results obtained.

From Equation 3.14, Q* = 862.24 units

From Equation 3.15, optimum shortage quantity, J* = 2.93 units

From Equation 3.16, optimum number of procurement cycles for the input item per production cycle, k* = 4.54

By observing Equation 3.8 and Equation 3.14, it can be said that the production lot size (by allowing shortages) is $\sqrt{(S+PF)/S}$ times the lot size when shortages are not allowed.

Similarly by observing Equation 3.9 and Equation 3.16, optimum frequency of ordering, k* (obtained as 4.54 in this example) is also $\sqrt{(S+PF)/S}$ times the frequency of ordering when shortages were not allowed.

Using Equation 3.21, optimum total annual cost, E* = Rs. 39,146.47.

Decrease in cost by allowing shortages = 39,155.19 – 39,146.47 = Rs. 8.72

By subtracting the right-hand side (R.H.S.) of Equation 3.21 from that of Equation 3.10, analytically, this decrease in cost,

$$\sqrt{\frac{2(yp-r)PFCR}{yp}} - \sqrt{\frac{2PFRCS(yp-r)}{yp(S+PF)}}$$

$$= \left[1 - \sqrt{\frac{S}{(S+PF)}}\right]\sqrt{\frac{2(yp-r)PFCR}{yp}}$$

3.4 PARTIAL BACKORDERING

This is a situation when a fraction b (b < 1) of shortage quantity is not backordered. Referring to Section 2.2.2, Equation 2.43 is developed as the cost concerning produced item i.e.,

$$= \frac{pJ^2(S+PF)}{2(yp-r)Q} + \frac{(yp-r)PFQ}{2p} - PFJ + \left[\frac{R}{y} - \frac{bpJ^2}{2(yp-r)Q}\right]P + \frac{RC}{yQ} - \frac{bpC}{2(yp-r)}\left[\frac{J}{Q}\right]^2 \tag{3.22}$$

Where P includes the value addition and S is the annual shortage cost per unit.

Annual cost concerning the input item is expressed by Equation 3.12, which is adjusted for the present case as,

$$= \frac{kA}{Q}\left[\frac{R}{y} - \frac{bJ^2p}{2Q(yp-r)}\right] + \frac{nQrGF}{2kyp} \tag{3.23}$$

Adding Equation 3.22 and Equation 3.23, total relevant cost,

$$E = \frac{RP}{y} + (C+kA)\frac{R}{yQ} - \frac{(C+kA)bJ^2p}{2Q^2(yp-r)} - \frac{bpJ^2P}{2(yp-r)Q} + \frac{(yp-r)PFQ}{2p} - PFJ + \frac{pJ^2(S+PF)}{2(yp-r)Q} + \frac{nQrGF}{2kyp} \tag{3.24}$$

As discussed in Section 2.2.2 and in Example 2.6, change in the frequency of setup and ordering due to partial backordering is negligible. Therefore, the third term of Equation 3.24 may be omitted and the equation reduces to:

$$E = \frac{RP}{y} + \frac{CR}{yQ} + \frac{kAR}{yQ} + \frac{(S+PF-bP)pJ^2}{2(yp-r)Q} + \frac{(yp-r)PFQ}{2p} - PFJ + \frac{nQrGF}{2kyp} \tag{3.25}$$

Now $\frac{\partial E}{\partial k} = 0$ shows, $k = Q\sqrt{\frac{nrGF}{2ARp}}$ (3.26)

$$\frac{\partial E}{\partial J} = 0 \text{ shows, } J = \frac{PF(yp-r)Q}{p(S+PF-bP)} \tag{3.27}$$

and ∂E/∂Q = 0 gives,

$$(C+kA)\frac{R}{yQ^2} + \frac{(S+PF-bP)pJ^2}{2(yp-r)Q^2} = \frac{(yp-r)PF}{2p} + \frac{nrGF}{2kyp}$$

Substituting Equation 3.26 and Equation 3.27,

$$Q^* = \sqrt{\frac{2pCR(S+PF-bP)}{yPF(yp-r)(S-bP)}} \tag{3.28}$$

From Equation 3.26, $k^* = \sqrt{\frac{nrGC(S+PF-bP)}{yPA(yp-r)(S-bP)}}$ (3.29)

and from Equation3.27, $J^* = \sqrt{\frac{2PFCR(yp-r)}{yp(S-bP)(S+PF-bP)}}$ (3.30)

Conditions for optimality, Equations 3.17–3.19, are satisfied with the feasibility condition S > P, otherwise also, optimal results obtained are feasible only when S > bP or simply S > P. This condition is easily satisfied in the real world, as the annual shortage costs are more than the production/procurement costs.

Substituting optimal results in Equation 3.25, optimal total relevant cost:

$$E^* = \frac{RP}{y} + \frac{1}{y}\sqrt{\frac{2nrGFAR}{p}} + \sqrt{\frac{2RCPF(yp-r)(S-bP)}{py(S+PF-bP)}} \tag{3.31}$$

Example 3.3

Show how decrease in the cost (by using optimal values) is observed when partial backordering exists, and comment.

Subtracting Equation 3.31 from Equation 3.21, decrease in cost,

$$= \sqrt{\frac{2PFRC(yp-r)}{py}}\left[\sqrt{\frac{S}{(S+PF)}} - \sqrt{\frac{(S-bP)}{(S+PF-bP)}}\right]$$

$$= \sqrt{\frac{2PFRC(yp-r)}{py}} \left[\frac{\sqrt{S(S+PF-bP)} - \sqrt{S(S+PF-bP)-bP^2F}}{\sqrt{(S+PF)(S+PF-bP)}}\right] \tag{3.32}$$

Let $X = S(S+PF-bP)$, and as $\left[\sqrt{X} - \sqrt{X-bP^2F}\right]$ is positive, decrease in the cost will be observed. This decrease in cost is advantageous if the lost customers are replaced by new customers without any effort. If appreciable efforts are needed, such as advertising, etc., then there is tradeoff between these costs and savings obtained by Equation 3.32. Estimation of shortage cost is difficult. However, if the costs of replacing the lost customers are included in the estimated shortage cost, then the net gain represented by Equation 3.32 can be achieved in the present context with the implementation of optimal results.

Example 3.4

Let, R = 1,200, C = Rs. 200, F = 0.2, r = 100, p = 120, y = 0.95, G = Rs. 10, n = 2, P = Rs. 30, A = Rs. 50, S = Rs. 50 and b = 0.3

Provide the optimal solution and show the effect of increase in shortage cost, S from the present level.

Using Equations 3.28–3.31,

Q* = 909.63 units
K* = 4.794
J* = 13.55 units
E* = Rs. 39,115.94

Now shortage cost S is increased by certain percentage from its present value of Rs. 50 and output values are obtained as follows:

S (Rs.)	% Increase in S	Q*	k*	J*	E*
60	20%	898.17	4.734	11.03	39,123.02
70	40%	890.39	4.693	9.30	39,127.94
80	60%	884.76	4.663	8.04	39,131.55
90	80%	880.49	4.640	7.08	39,134.32
100	100%	877.15	4.623	6.33	39,136.51

Considering the optimal results obtained, Q*, k* and J* decrease with increase in shortage cost. But total optimal cost E* increases with increase in shortage. Their sensitivity is computed with reference to optimal results obtained with S = Rs. 50, as follows:

S. No.	% Increase in S	% Decrease in Q*	% Decrease in k*	% Decrease in J*	% Increase in E*
1	20	**1.25**	1.25	18.59	0.018
2	40	2.11	2.11	31.36	0.031
3	60	2.73	2.73	40.66	0.039
4	80	3.20	3.20	47.75	0.047
5	100	3.57	3.57	53.28	0.053

Percentage decrease in Q* and k* are similar. This is because k* is directly proportional to Q* with reference to change in shortage cost, as evident from Equation 3.26. Considering the difference between values at S. No. (N) and S. No. (N–1) in each column, it can be observed that the sensitivity of output parameter values decrease with increase in shortage cost. This conclusion is because of constant percentage increase in S (i.e. of 20%). For example, difference in the values of column concerning % decrease in J* are 31.36 – 18.59 = 12.77, 40.66 – 31.36 = 9.3, 47.75 – 40.66 = 7.09 and 53.28 – 47.75 = 5.53. Therefore these values are 12.77, 9.3, 7.09 and 5.53, respectively, which show decrease trend—and hence, sensitivity decreases with increase in shortage cost.

In the present example, percentage increase in E* seems to be less significant. This depends on values of input parameters. Refer Equation 3.31. The first two components of the total cost are concerned with value added production cost and procurement of the input item, which do not depend on shortage costs. The third component relates to change in shortage cost. Considering the input parameters, a major portion of the total cost is constant with respect to change in shortage cost, and therefore, percentage increase in E* is less. It is also dependent on how much value addition is made by the manufacturing organization.

At present, only one parameter, i.e. shortage cost, is considered for sensitivity analysis. Similar procedures may be conducted for other parameters of concern to the management, but the sensitivity with respect to the total system cost (whatever significant or insignificant it may be) provides a lot of insights to the management. This gives the information regarding how much any change in any input parameter contributes to the total system cost. For instance, depending on the economic scenario, annual carrying cost fraction changes, and the mathematical modelling of the real system helps in the decision making from time to time.

3.4.1 Finite Replenishment Rate of the Input Item

It replenishment rate of the input item is finite, i.e. d, then total cost:

$$
\begin{aligned}
E = \frac{RP}{y} + (C + kA)\frac{R}{yQ} - \frac{(C + kA)bJ^2 p}{2Q^2(yp - r)} - \frac{bpJ^2 P}{2(yp - r)Q} \\
+ \frac{(yp - r)PFQ}{2p} - PFJ + \frac{pJ^2(S + PF)}{2(yp - r)Q} + \frac{(d - np)nrGFQ}{2kdyp}
\end{aligned}
\tag{3.33}
$$

Following the similar procedure, optimal results are obtained as follows along with the feasibility condition S > P:

$$k^* = \sqrt{\frac{nrGC(d-np)(S+PF-bP)}{yPAd(yp-r)(S-bP)}} \tag{3.34}$$

Q* and J* similar to Equation 3.28 and Equation 3.30, respectively. Substituting optimal values in Equation 3.33, total relevant cost is obtained.

In the present formulations so far, a single input item has been considered. If n is equal to 1, then these are also suitable for several input items, which is a general case. For example, in order to produce 1 m of 150 mm class 15 asbestos cement pressure pipe, approximately 2.14 kg asbestos fibre and 13.16 kg cement are required. In such case, a unit of the input item may be assumed as 13.16 kg cement and 2.14 kg asbestos fibre, and the joint ordering cost for both the input items may be estimated. However, the next section deals with multiple input items in an explicit manner.

3.5 MULTIPLE INPUT ITEMS

Let:

Number of input items required = m
Fixed ordering cost for the input item i = A_i
Frequency of ordering for the input item i in a production cycle = k_i
Units of the input item required per unit end item = n_i
Unit purchase cost of the input item i = G_i
Replenishmentrate of the input item i in units per period = d_i

Figure 3.4 shows the procurement of an item i in smaller lot sizes during production time Q/p.

Annual cost concerning the input item was developed in Section 3.1 expressed by Equation 3.2:

$$= \frac{RkA}{yQ} + \frac{(d-np)nQrGF}{2kdyp}$$

As there are m number of input items, the preceding equation can be written as,

$$\text{Total cost for multiple input items} = \sum_{i=1}^{m} \frac{R}{yQ}(k_i A_i) + \sum_{i=1}^{m} \frac{(d_i - pn_i)n_i QrG_i F}{2k_i d_i yp} \tag{3.35}$$

As expressed by Equation 3.1,

$$\text{Total cost concerning the end product} = \frac{(yp-r)PFQ}{2p} + \frac{RC}{yQ} + \frac{PP}{y} \tag{3.36}$$

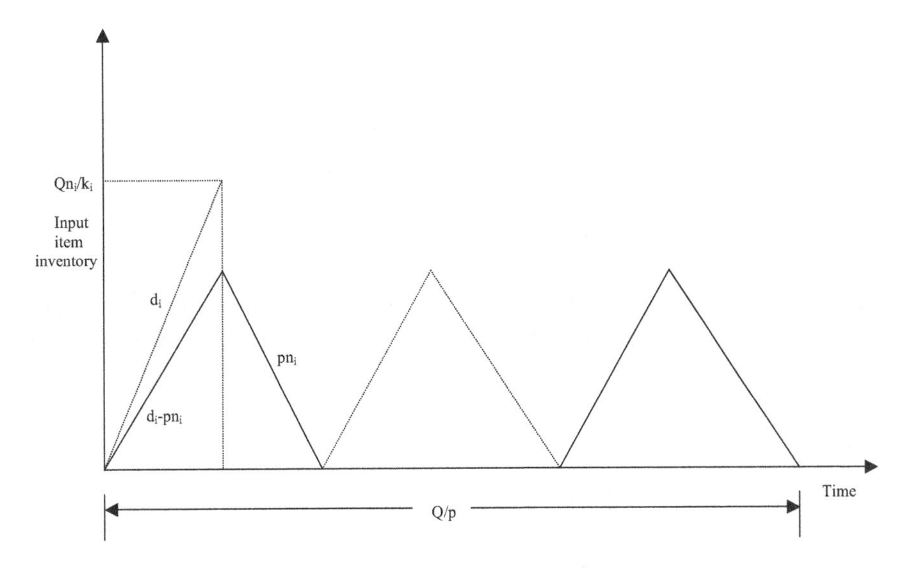

FIGURE 3.4 Procurement of an item i in the context of several input items.

Adding Equation 3.35 and Equation 3.36, total annual cost:

$$E = \frac{RP}{y} + \frac{RC}{yQ} + \frac{(yp - r)PFQ}{2p} + \frac{R}{yQ}\sum_{i=1}^{m} k_i A_i + \sum_{i=1}^{m} \frac{(d_i - pn_i)n_i QrG_i F}{2k_i d_i yp} \tag{3.37}$$

As Equation 3.37 is a convex function in terms of Q and each k_i, $\partial E/\partial Q = 0$ shows:

$$Q = \sqrt{\frac{2pR\left(C + \sum_{i=1}^{m} k_i A_i\right)}{Fy\left[P(yp - r) + (r/y)\sum_{i=1}^{m}(d_i - pn_i)n_i G_i/(k_i d_i)\right]}} \tag{3.38}$$

Q from Equation 3.38 is substituted in Equation 3.37 in order to obtain the total cost, E in terms of k_i, as

$$E = (RP)/y + \sqrt{\left(\frac{2FR}{py}\right)\left(C + \sum_{i=1}^{m} k_i A_i\right)\left[P(yp - r) + \frac{r}{y}\sum_{i=1}^{m}\frac{(d_i - pn_i)n_i G_i}{k_i d_i}\right]} \tag{3.39}$$

Now equating partial derivative of E with respect to each k_i to zero, shows:

$$k_i = \sqrt{\frac{\left(C + \sum_{i=1}^{m} k_i A_i\right) r(d_i - pn_i)n_i G_i}{yA_i d_i \left[P(yp - r) + (r/y)\sum_{i=1}^{m}(d_i - pn_i)n_i G_i / (k_i d_i)\right]}} \tag{3.40}$$

Using Equation 3.40 for k_i, i=1, 2, . . . m, optimal k_i are obtained, and by substituting them into Equation 3.39, total cost is evaluated. This is illustrated with the following example.

Example 3.5

The data concerning finished product are as follows:

Annual demand, R = 1,200
Setup cost, C = Rs. 200
Proportion of nondefective items in a lot, y = 0.95
Demand rate per period, r = 100
Production rate per period, p = 120
Production cost per unit (including value addition), P = Rs. 30
In addition, the annual carrying cost fraction, F = 0.2

Assume that two number of input items are needed in order to manufacture the end item, i.e. m = 2. A_i, G_i, n_i and d_i for i = 1, 2 are as follows:

A_1 = Rs. 20, A_2 = Rs. 30

G_1 = Rs. 10, G_2 = Rs. 5

$n_1 = 1$, $n_2 = 2$

$d_1 = 150$, $d = 300$

Now $(d_1 - pn_1)n_1G_1 = 300$,
$(d_2 - pn_2)n_2G_2 = 600$,
and $(yp - r) = 14$

From Equation 3.40,
$$k_1 = \sqrt{\frac{(200 + 20k_1 + 30k_2) \times 100 \times 300}{3000\left[399 + 100\{(2/k_1) + (2/k_2)\}\right]}} \tag{3.41}$$

$$\text{and } k_2 = \sqrt{\frac{(200+20k_1+30k_2)\times 100\times 600}{9000\left[399+100\{(2/k_1)+(2/k_2)\}\right]}} \tag{3.42}$$

Dividing Equation 3.42 by Equation 3.41,

$$\frac{k_2}{k_1} = \sqrt{\frac{2}{3}}$$

$$\text{Or } k_2 = k_1\sqrt{2/3}$$

Substituting k_2 in Equation 3.41,

$$k_1^{\,2} = \frac{(200+20k_1+30k_1\sqrt{2/3})\times 100\times 300}{3000\left[399+100\left\{(2/k_1)+\left(\sqrt{6}/k_1\right)\right\}\right]}$$

$$\text{Or } 399k_1^{\,2} = 2000$$

$$\text{Or } k_1 = 2.24$$

$$\text{and as } k_2 = k_1\sqrt{2/3}\text{ , } k_2 = 1.83$$

Substituting k_i, i = 1, 2 in Equation 3.39, total cost, E = 37,894.74 + 890.93 = Rs. 38,785.67.

Where RP/y = Rs. 37,894.74, which is a common component of the cost as long as y is constant. It should essentially be included while conducting sensitivity analysis with respect to quality level, as explained in Chapter 2. However, for convenience, this component of the cost may be excluded in order to obtain integer values of k_1 and k_2 using integer programming. To implement the solutions, frequency of ordering k_1 and k_2 should be integer. The procedure to obtain integer optimum is shown in Figure 3.5. Node I represents non-integer solution which is recently obtained. Arbitrarily selecting k_1 for further branching and because k_1 = 2.24, the nearest integer values are 2 and 3; therefore, Node II and Node III correspond to k_1 = 2 and 3, respectively. At Node II, with k_1 = 2 and k_2 = 1.83, cost is obtained as Rs. 891.66. Similarly at Node III, computations are made. From Node II, branches emanate with integer k_2 as 1 and 2, reaching to Node IV and Node V, respectively, and costs are computed. Similarly, from Node III, calculations are made for Node VI and Node VII, as shown in Figure 3.5. The lowest cost, i.e. Rs. 892.44, is represented by Node V with integer optimum values of k_1 and k_2 as 2 each. Now optimal solutions are as follows:

$k_1{}^* = 2$
$k_2{}^* = 2$
E^* = Rs. 892.44 + Rs. 37,894.74 = Rs. 38,787.18

From Equation 3.38, using $k_1{}^*$, Q^* = 849.24 units.

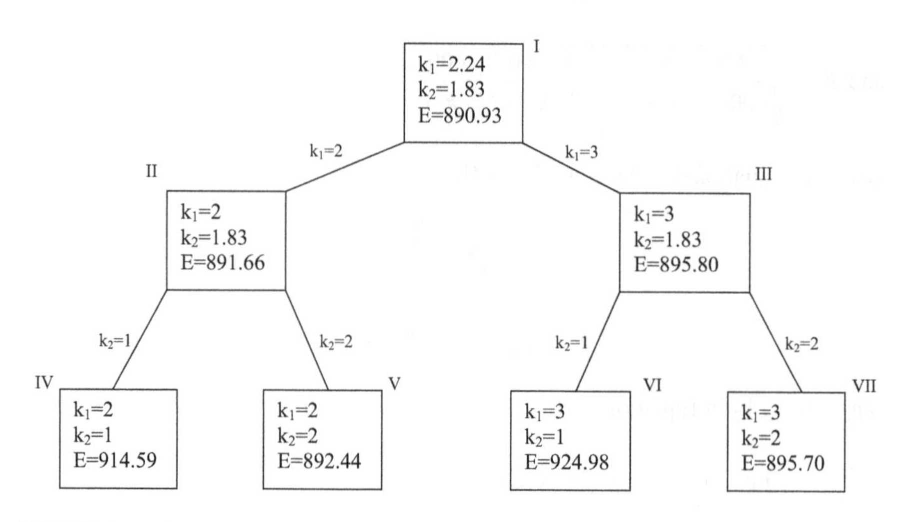

FIGURE 3.5 Obtaining integer optimum k_i.

3.5.1 Generalized Procedure to Evaluate K_i

In Example 3.5, k_i, i = 1, 2 were evaluated as 2.24 and 1.83 using Equation 3.40. If the number of input items is greater, then computational difficulties arise. The following general procedure is evolved to evaluate k_i:

$$\text{Let } X = \sqrt{\frac{\left(C+\sum_{i=1}^{m} k_i A_i\right) r}{y\left[P(yp-r)+(r/y)\sum_{i=1}^{m}(d_i - pn_i)n_i G_i/(k_i d_i)\right]}} \tag{3.43}$$

$$\text{Now, from Equation 3.40, } k_i = X\sqrt{\frac{n_i G_i (d_i - pn_i)}{A_i d_i}}, i = 1,2,...m \tag{3.44}$$

$$\text{Or } X^2 = \frac{k_i^2 A_i d_i}{n_i G_i (d_i - pn_i)} \tag{3.45}$$

Now from Equation 3.43:

$$X^2 yP(yp-r) + X^2 r\sum_{i=1}^{m}(d_i - pn_i)n_i G_i/(k_i d_i) = rC + r\sum_{i=1}^{m} k_i A_i \tag{3.46}$$

From Equation 3.44, $\frac{n_i G_i (d_i - pn_i)}{d_i} = \frac{A_i k_i^2}{X^2}$, and substituting it in Equation 3.46,

$$X^2 yP(yp-r) + r\sum_{i=1}^{m} A_i k_i = rC + r\sum_{i=1}^{m} A_i k_i$$

$$\text{Or } X^2 yP(yp - r) = rC$$

Substituting X^2 from Equation 3.45, optimal value of k_i is obtained as:

$$k_i^* = \sqrt{\frac{rCn_i G_i (d_i - pn_i)}{A_i d_i yP(yp - r)}}, i = 1, 2, ...m \tag{3.47}$$

3.5.2 Instantaneous Procurement

If procurement of multiple input items is instantaneous, i.e. infinite replenishment rate, then the optimal results are obtained as follows:

$$k_i^* = \sqrt{\frac{rCn_i G_i}{A_i yP(yp - r)}}, i = 1, 2, ...m \tag{3.48}$$

$$E^* = \frac{RP}{y} + \sqrt{\left(\frac{2FR}{py}\right)\left(C + \sum_{i=1}^{m} k_i A_i\right)\left[P(yp - r) + (r / y)\sum_{i=1}^{m} n_i G_i / k_i^*\right]} \tag{3.49}$$

$$\text{and } Q^* = \sqrt{\frac{2pR\left(C + \sum_{i=1}^{m} k_i A_i\right)}{Fy\left[P(yp - r) + (r / y)\sum_{i=1}^{m} n_i G_i / k_i^*\right]}} \tag{3.50}$$

3.6 FREQUENT DELIVERY OF PRODUCED ITEMS

In the context of supply chain, a manufacturing company may need to deliver a small quantity of produced items frequently to another purchasing firm. This situation is shown in Figure 3.6, along with instantaneous procurement of the input item.

Let:

m= time between successive deliveries of the produced item to another purchasing firm.
w= number of deliveries during up time, i.e. Q/p
p= production rate per year
r= specified small quantity of the produced item to be delivered periodically to the buyer
I= number of deliveries of produced item during the cycle (cycle time = yQ/R)
x = number of deliveries during down time

An area under inventory cycle diagram concerning produced item inventory will now be needed to evaluate average inventory and then carrying cost for the produced

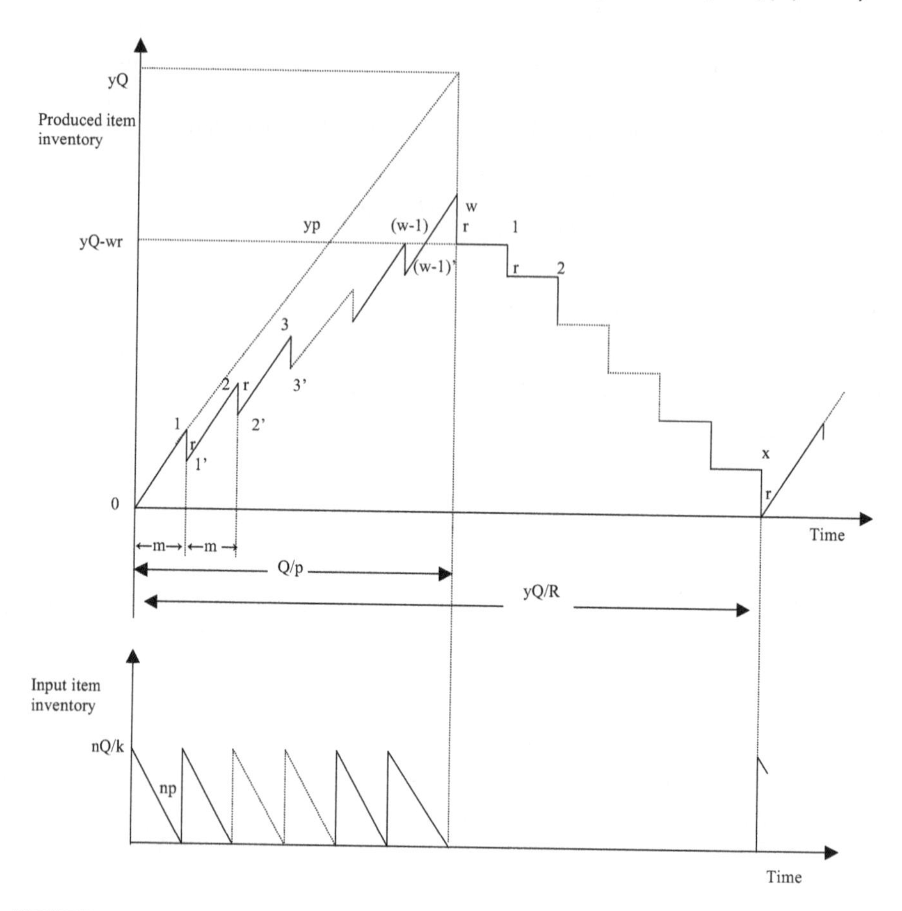

FIGURE 3.6 Production-inventory cycle.

items. Area under the inventory cycle diagram is obtained by adding area during up time and area during down time.

Area under 0 – 1 = ½. m. ypm
Area under 1' – 2 = ½. m. ypm + m (ypm – r)
Area under 2' – 3 = ½. m. ypm + m. 2 (ypm – r)
Area under (w – 1)'—w = ½. m. ypm + m (w – 1) (ypm-r)
Area during up time = ½. m. ypm w + [1 + 2 + 3 + . . . + (w – 1)] m (ypm – r)

$$= \frac{mw}{2} \cdot ypm + \left[\frac{(w-1)}{2}\right][2+(w-2)]m(ypm-r)$$

$$= (mw/2)[ypm+(w-1)(ypm-r)] \tag{3.51}$$

As there are x number of deliveries during down time, area during down time = m(yQ – wr) + m(yQ – wr – r) + . . .

$$+ [yQ - wr - (x-1)r]$$

$$= mx(yQ - wr) - \frac{mrx(x-1)}{2} \tag{3.52}$$

Adding Equation 3.51 and Equation 3.52, area under inventory cycle diagram,

$$= mx(yQ - wr) - \frac{mrx(x-1)}{2} + \frac{mw}{2}\left[ypm + (w-1)(ypm - r)\right] \tag{3.53}$$

Average inventory is obtained by dividing Equation 3.53 by (yQ/R), i.e., the cycle time (mI) and substituting:

m = r/R

I = yQ/r

And $w = \frac{Q}{pm} = \frac{QR}{pr}$

$$x = I - w = \frac{yQ}{r} - \frac{QR}{pr}$$

Average inventory $= \frac{1}{2}\left[r + \frac{(yp - R)Q}{p}\right]$

And annual inventory carrying cost for produced item:

$$= \frac{1}{2}\left[r + \frac{(yp - R)Q}{p}\right]PF$$

Remaining cost components for the manufactured item as well as annual cost concerning the input item are obtained as discussed before. After developing the expression for total relevant cost, optimal parameters and optimal total cost are obtained following the usual procedure.

In order to minimize the overall costs and to become an efficient link in the supply chain, reduction of setup and ordering cost is usually proposed, for which an additional investment is needed. The model formulation is useful for tradeoff between the cost savings due to the said proposal and the additional investment.

Exercises

1. Consider the industrial problem such as shown in the following figure:

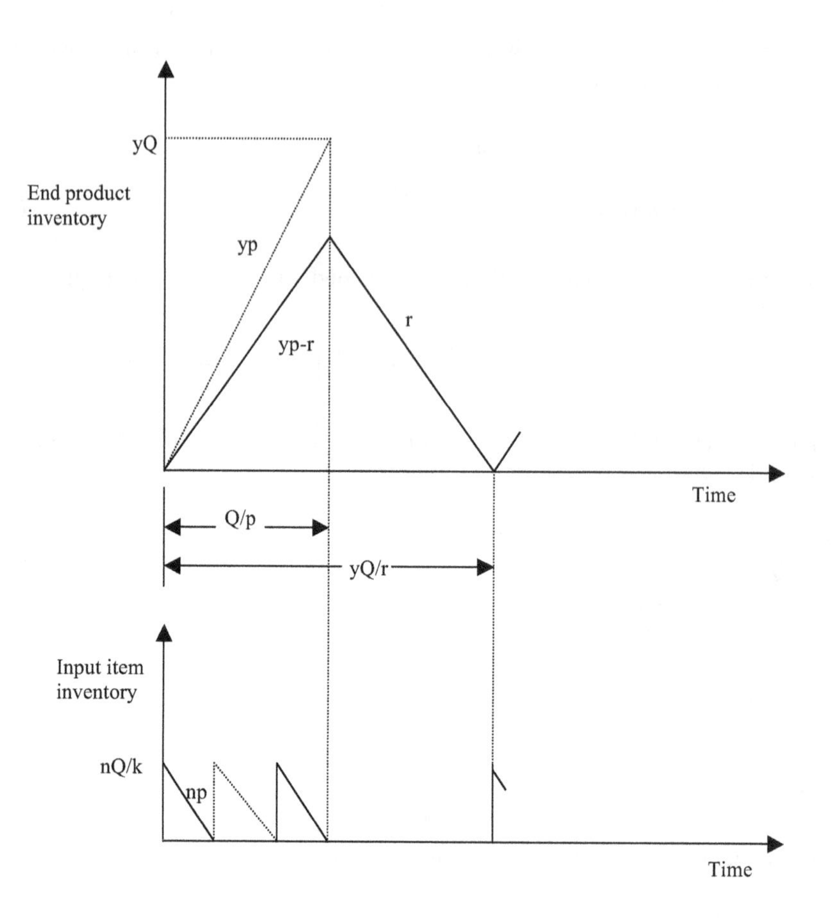

What will be your approach for modelling input item procurement along with the end product manufacture? Analyze this in a generalized form and obtain the optimum expressions for:

a. Production lot size
b. Number of orders for procurement of the input item in a production cycle
c. Total cost

2. Consider the following parameters:

Annual demand = 1,200
Production setup cost = Rs. 250
Annual inventory holding cost fraction = 0.25
Demand rate of end item per period = 100
Production rate per period = 130
Proportion of acceptable end items in a lot = 0.97
Unit purchase cost of the input item = Rs. 11
Units of the input item required per unit of end item = 3
Unit production cost (including value addition) = Rs. 35

Fixed ordering cost for the input item = Rs. 60
Obtain the optimum values of production lot size, frequency of ordering of the input item and total cost.

Also conduct the sensitivity analysis concerning the following parameters:

a. Fixed ordering cost for the input item
b. Proportion of acceptable end items in a lot
c. Fixed production setup cost for the end product

3. In Exercise 1, consider that the replenishment of the input item is as follows instead of instantaneous procurement:

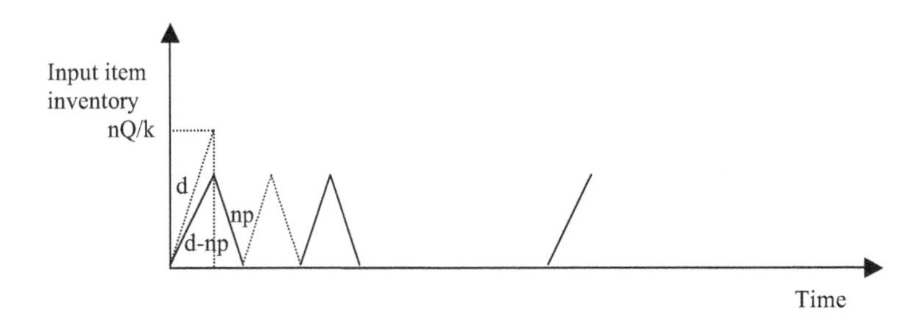

Analyze this in a generalized form (along with the end product manufacture) and obtain the optimum expressions for:

a. Production lot size
b. Number of orders for procurement of the input item in a production cycle
c. Total cost

4. Consider the following parameters:

Fixed ordering cost for the input item = Rs. 55
Unit purchase cost of the input item = Rs. 12
Annual demand = 1,200
Replenishment rate of the input item per period = 400
Production setup cost = Rs. 200
Annual inventory holding cost fraction = 0.3
Demand rate of end item per period = 100
Production rate per period = 130
Proportion of nondefective end items in a lot = 0.97
Units of the input item required per unit of end item = 3
Unit production cost (including value addition) = Rs. 30

Obtain the optimum values of production lot size, frequency of ordering of the input item and total cost.

Also conduct the sensitivity analysis concerning the following parameters:

a. Proportion of nondefective end items in a lot

b. Fixed ordering cost for the input item
c. Fixed production setup cost for the end product

5. Consider the problem as shown below:

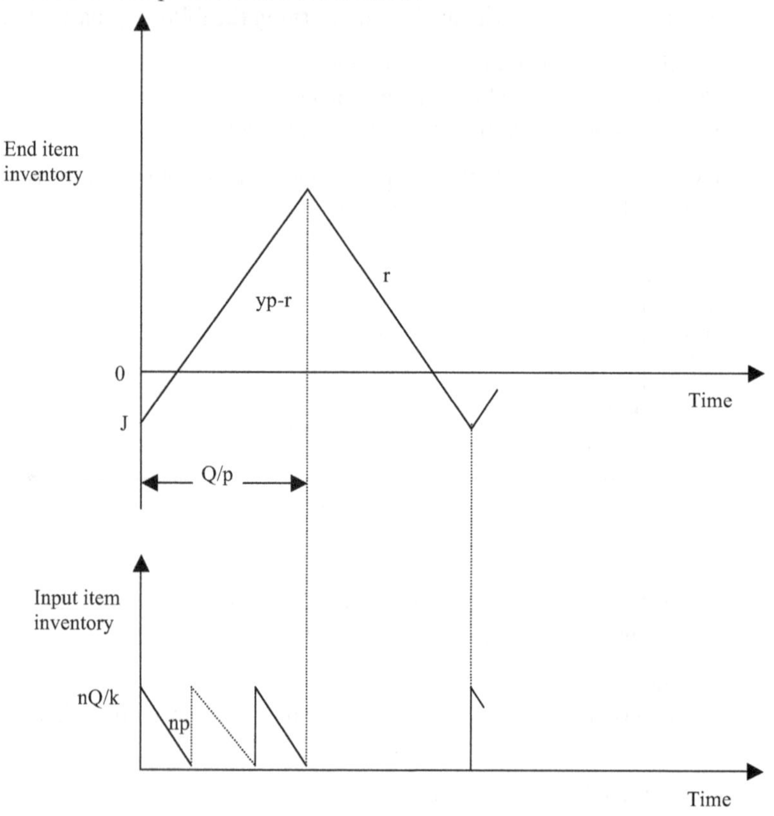

What will be your approach (particularly because it is a three variables optimization problem) for modelling the input item procurement, along with the end product manufacture, with an inclusion of shortages? Analyze this in a generalized form and obtain the optimum expressions for:

a. Production lot size
b. Maximum relevant shortage quantity
c. Number of orders for procurement of the input item in a production cycle
d. Total cost

6. Assume the input data as follows:

Production setup cost = Rs. 225
Annual inventory holding cost fraction = 0.24
Demand rate of end item per period = 100
Production rate per period = 125
Proportion of acceptable end items in a lot = 0.99

Annual demand = 1200
Unit purchase cost of the input item = Rs. 13
Units of the input item required per unit of end item = 2
Unit production cost (including value addition) = Rs. 32
Annual shortage cost per unit end product = Rs. 150
Fixed ordering cost for the input item = Rs. 65

Obtain the optimum values of:

a. Production lot size
b. Maximum shortage quantity
c. Frequency of ordering of the input item
d. Total cost

Also, conduct sensitivity analysis concerning the following parameters:

a. Fixed ordering cost for the input item
b. Fixed production setup cost for the end product
c. Annual shortage cost per unit of the end product
d. Proportion of acceptable end items in a lot

7. In Exercise 5, implement the finite replenishment of the input item (as shown in the following figure) instead of instantaneous procurement:

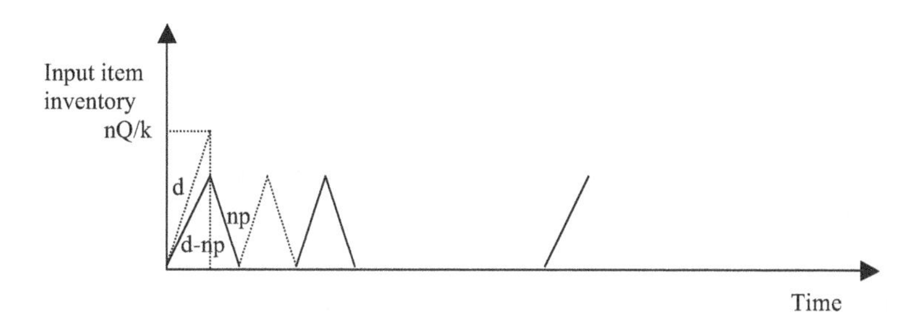

Analyze this in a generalized form (along with end product manufacture) and obtain the optimum expressions for:

a. Production lot size
b. Number of orders for procurement of the input item in a production cycle
c. Maximum relevant shortage quantity
d. Total cost

8. Following details are provided:

Unit production cost (including value addition) = Rs. 33
Annual shortage cost per unit of the end product = Rs. 165
Fixed ordering cost for the input item = Rs. 70
Production setup cost = Rs. 235

Annual inventory holding cost fraction = 0.23
Demand rate of end item per period = 100
Production rate per period = 115
Replenishment rate of the input item per period = 400
Proportion of nondefective end items in a lot = 0.96
Annual demand = 1,200
Unit purchase cost of the input item = Rs. 14
Units of the input item required per unit of end item = 3

Obtain the optimum values of:

a. Production lot size
b. Maximum shortage quantity
c. Frequency of ordering of the input item
d. Total cost

Also conduct the sensitivity analysis concerning the following parameters:

a. Proportion of nondefective end items in a lot
b. Fixed ordering cost for the input item
c. Fixed production setup cost for the end product
d. Annual shortage cost per unit of the end product

9. What do you understand by partial backordering in manufacturing system? Incorporate this in the manufacturing of end item along with the purchase of raw material where replenishment rate is:

   a. Infinite
   b. Finite

   Also, get the generalized expressions for the optimized values of output parameters such as:

   a. Manufacturing batch size
   b. Frequency of ordering for purchase of raw material in a manufacturing cycle
   c. Relevant shortage quantity
   d. Total relevant cost

10. Following are the input details with infinite replenishment rate of raw material:

    Fraction of shortage quantity which is not backordered = 0.3
    Production setup cost = Rs. 215
    Annual inventory holding cost fraction = 0.27
    Demand rate of end item per period = 100
    Production rate per period = 125
    Proportion of acceptable end items in a lot = 0.99
    Annual demand = 1,200
    Unit purchase cost of the input item = Rs. 9
    Units of the input item required per unit of end item = 2

Unit production cost (including value addition) = Rs. 32
Annual shortage cost per unit of the end product = Rs. 120
Fixed ordering cost for the input item = Rs. 65

Obtain the optimum values of:

a. Production lot size
b. Maximum shortage quantity
c. Frequency of ordering of the input item
d. Total cost

Also conduct the sensitivity analysis concerning the following parameters:

a. Fraction of shortage quantity which is not backordered
b. Fixed ordering cost for the input item
c. Fixed production setup cost for the end product
d. Annual shortage cost per unit of the end product
e. Proportion of acceptable end items in a lot

11. Following are the input details with finite replenishment rate of raw material:

Unit production cost (including value addition) = Rs. 35
Annual shortage cost per unit of the end product = Rs. 135
Fixed ordering cost for the input item = Rs. 70
Production setup cost = Rs. 205
Annual inventory holding cost fraction = 0.29
Demand rate of end item per period = 100
Production rate per period = 120
Replenishment rate of the input item per period = 450
Proportion of nondefective end items in a lot = 0.96
Annual demand = 1200
Unit purchase cost of the input item = Rs. 10
Units of the input item required per unit of the end item = 3

Obtain the optimum values of:

a. Frequency of ordering of the input item
b. Production lot size
c. Maximum shortage quantity
d. Total cost

Also conduct the sensitivity analysis concerning the following parameters:

a. Proportion of nondefective end items in a lot
b. Fixed ordering cost for the input item
c. Fraction of shortage quantity which is not backordered
d. Fixed production setup cost for the end product
e. Annual shortage cost per unit of the end product

12. Analyze the scenario concerning procurement of several input items (as shown in the following figure) related to an end product:

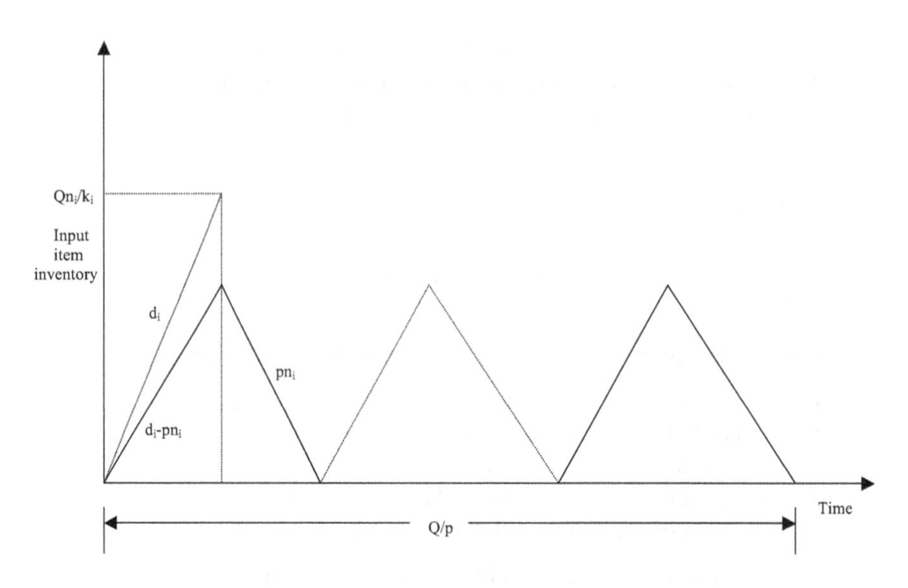

13. The data concerning the end product are as follows:

    Setup cost = Rs. 240
    Proportion of nondefective items in a lot = 0.97
    Demand rate per period = 100
    Production rate per period = 130
    Production cost per unit (including value addition) = Rs. 35
    Annual demand = 1,200

    Furthermore, the two numbers of input items are required for manufacturing the end product. The parameters for input item 1 and input item 2 are given below:

    Fixed ordering cost for input item 1 = Rs. 35
    Fixed ordering cost for input item 2 = Rs. 25
    Units of input item 1 required per unit end item = 2
    Units of input item 2 required per unit end item = 1
    Unit purchase cost of input item 1 = Rs. 8
    Unit purchase cost of input item 2 = Rs. 12
    Replenishment rate of input item 1 in units per period = 300
    Replenishment rate of input item 2 in units per period = 200

    a. Obtain the frequency of ordering for input item 1 and input item 2 in a manufacturing cycle assuming the annual carrying cost fraction = 0.25. Also calculate the total relevant cost.
    b. Conduct the sensitivity analysis related to the following parameters:

        i. Fixed ordering cost for the input items
        ii. Proportion of acceptable end items in a lot
        iii. Fixed production setup cost for the end product

c. Find out the integer optimum solution in terms of the frequency of ordering for the input items along with the corresponding total relevant cost. Explain the approach with the help of Fig. by creating various nodes starting from the solution obtained earlier.

14. If the number of input items is greater, then computational difficulties might arise. In order to overcome this, discuss a generalized procedure to evaluate frequency of ordering for the input items in a production cycle related to the following cases:

    a. Instantaneous procurement of the input items
    b. Finite replenishment rate of the input items

15. In some cases, a manufacturing company may need to dispatch small quantity of produced items (as shown in the following figure) frequently to the consumer company:

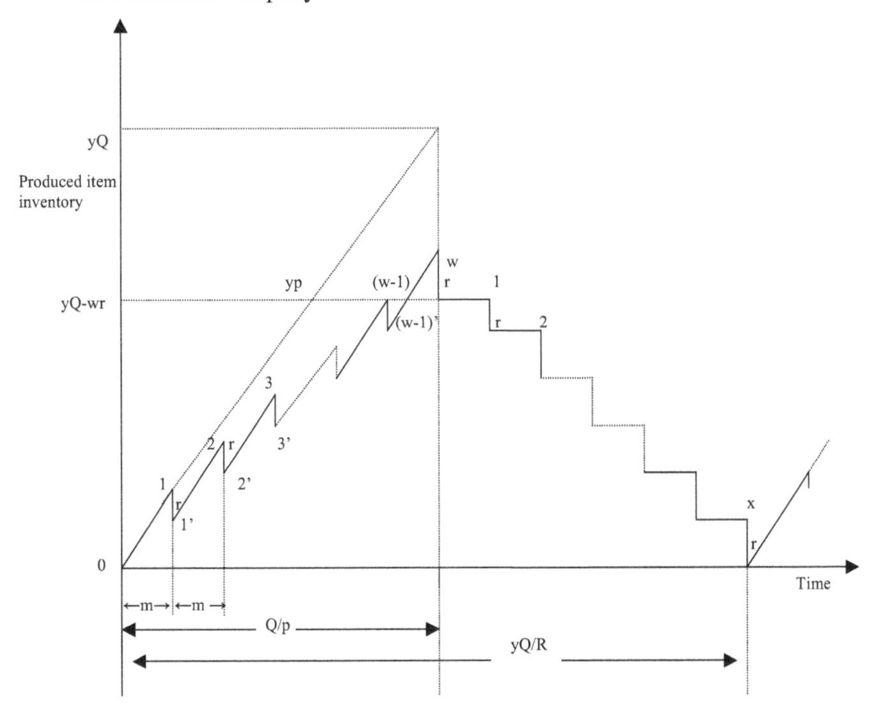

    a. Explain the approach for incorporating instantaneous procurement of the input item.
    b. Extend the approach for including finite replenishment rate of the input item.

4 Price Variation

Industrial or business organizations often face the following situations:

1. There is a price reduction concerning an item which may be purchased during a short period, i.e. the discount offer is applicable for a short duration. Usually, large quantities are procured during the availability of such an offer. A policy is to be developed to deal with short-term or temporary price reductions.
2. A price increase is declared concerning an item which will be effective after a certain time period. The organization may react by procuring large quantities of that item before the price increase becomes effective. With the help of modelling processes, appropriate decisions are made.

4.1 TEMPORARY PRICE REDUCTION

Let:

C = fixed ordering cost per order
P = unit purchase cost
p = replenishment rate, including defective units, in units per period
d = unit price reduction
Q_a = special order quantity

Figure 4.1 shows a finite replenishment situation. Temporary price discount is offered between time t_s and t_f. The offer of reduction in price will start at time t_s and it will end at time t_f. Larger quantities, i.e. special order quantity Qa, is procured to take the advantage of short-term price discount offer in which the unit price is (P – d) instead of P. Replenishment for special order Qa starts when stock position is q between time t_s and t_f, i.e. at time t_r.

The special order Q_a is consumed completely at time t_g, after which normal procurement starts. The situation between t_r and t_g is under consideration.

Without a Special Order

This is shown by dotted line in Figure 4.1. Referring to Section 2.2, optimal ordering quantity Q* is expressed by Equation 2.27 as

$$Q^* = \sqrt{\frac{2pRC}{(yp - r)PFy}} \tag{4.1}$$

DOI:10.1201/9781003213994-4

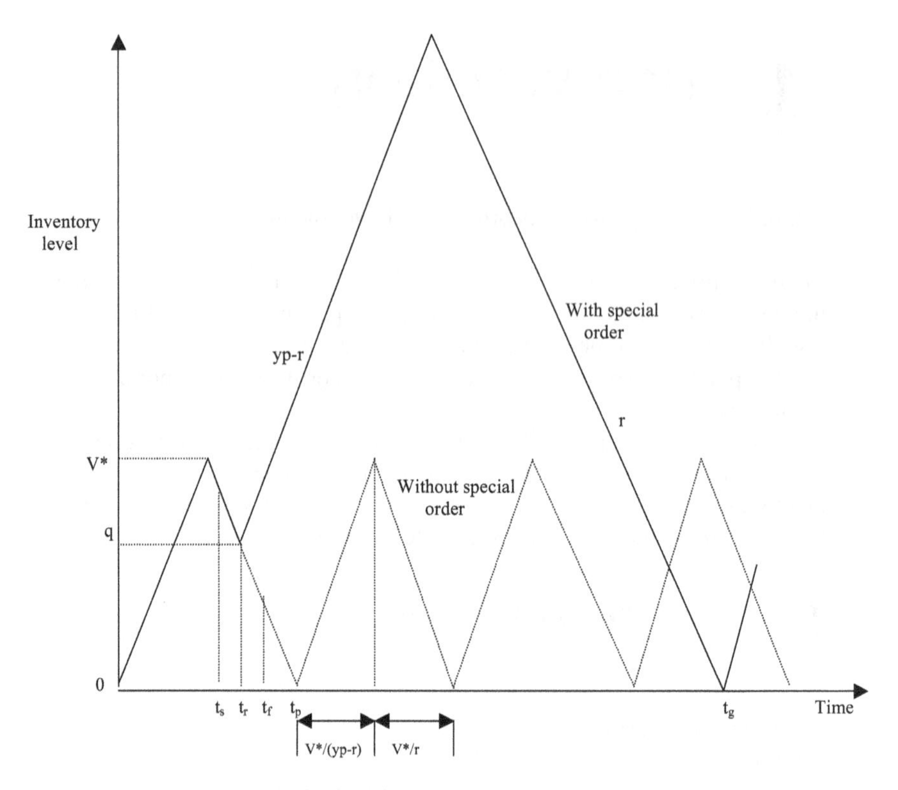

FIGURE 4.1 Transitional discounting situation.

and optimal maximum inventory during the cycle, $V^* = \dfrac{(yp-r)Q^*}{p}$ (4.2)

Number of orders placed during $t_r - t_g$ for the situation without special orders = Q_a/Q^*

$$\text{And ordering cost} = \frac{Q_a}{Q^*} \cdot C \tag{4.3}$$

During $t_r - t_g$, total quantity Q_a (although in lot size Q^*) is replenished and consumed, therefore:

$$\text{Purchase cost} = P.Q_a \tag{4.4}$$

Stock q is consumed during $t_r - t_p$ and then Q^* is ordered for number of cycles Q_a/Q^*.

Average inventory q/2 exists for (q/r) periods or (q/r); (r/R) = q/R year, and therefore, inventory carrying cost for period $t_r - t_p$ = (q/2)· (q/R)· PF = $(q^2PF)/2R$ (4.5)

During $t_p - t_g$, time for one cycle $= \dfrac{V^*}{(yp-r)} + \dfrac{V^*}{r}$

$$= \frac{ypV^*}{r(yp-r)} \text{ periods}$$

$$= \frac{ypV^*}{R(yp-r)} \text{ year} = \frac{yQ^*}{R},$$

substituting V* from Equation 4.2.

As the average inventory is (V*/2) and there are (Q_a/Q*) number of cycles for the period (t_p–t_g),

$$\text{Carrying cost} = \frac{V^*}{2} \cdot \frac{yQ^*}{R} \cdot \frac{Q_a}{Q^*} \cdot PF$$

Substituting V* from Equation 4.2,

$$\text{Carrying cost} = \frac{y(yp-r)Q^* Q_a PF}{2pR} \tag{4.6}$$

Adding Equation 4.5 and Equation 4.6, inventory carrying cost for the period under consideration ($t_r - t_g$):

$$= \frac{q^2 PF}{2R} + \frac{y(yp-r)Q^* Q_a PF}{2pR} \tag{4.7}$$

Adding Equation 4.3, Equation 4.4 and Equation 4.7, total cost without a special order,

$$T = PQ_a + \frac{Q_a}{Q^*} \cdot C + \frac{q^2 PF}{2R} + \frac{y(yp-r)Q^* Q_a PF}{2pR} \tag{4.8}$$

With a Special Order

Since only one special order is placed with ordering quantity Q_a, which is procured with price per unit as (P – d),

$$\text{Purchase cost} = Q_a\,(P - d) \tag{4.9}$$

$$\text{And ordering cost} = C \tag{4.10}$$

In order to evaluate inventory carrying cost between time t_r to t_g, relevant portion of Figure 4.1 is shown following:

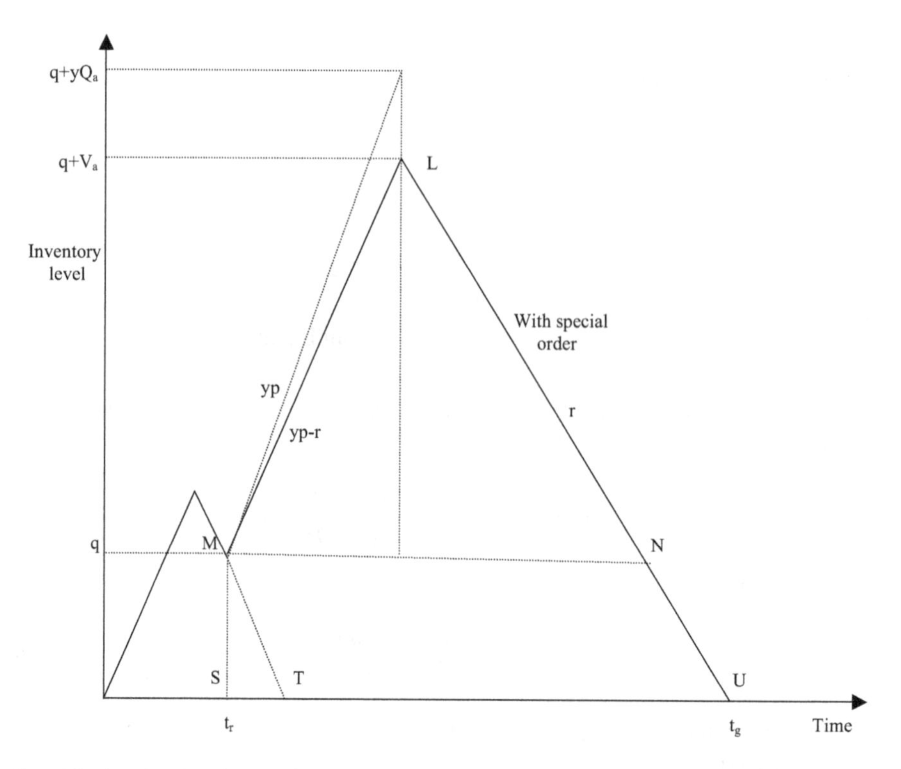

Considering L-M-N in the figure,

$$\frac{V_a}{(yp-r)} = \frac{yQ_a}{yp}$$

$$\text{Or } V_a = \frac{(yp-r)Q_a}{p}$$

As M-N in the figure is yQ_a/R year, carrying cost

$$= \frac{V_a}{2} \cdot \frac{yQ_a(P-d)F}{R}$$

$$= \frac{y(yp-r)Q_a^2(P-d)F}{2pR} \tag{4.11}$$

Consider M-S-T in the figure, q procured at unit price P is consumed and carrying cost

$$= (q/2) \cdot (q/R) \cdot PF = (q^2PF)/(2R) \tag{4.12}$$

For the remaining portion M-T-U-N in the figure, q (which is a part of Q_a) procured at unit price (P – d) exists at constant level for the purpose of inventory valuation and therefore,

$$\text{Carrying cost } = q \cdot \frac{yQ_a(P-d)F}{R} \tag{4.13}$$

Adding Equations 4.11–4.13, inventory carrying cost

$$= \frac{y(yp-r)Q_a^2(P-d)F}{2pR} + \frac{q^2PF}{2R} + \frac{yq(P-d)FQ_a}{R} \tag{4.14}$$

Adding Equation 4.9, Equation 4.10 and Equation 4.14, total cost with a special order,

$$T_a = C + (P-d)Q_a + \frac{q^2PF}{2R} + \frac{yq(P-d)FQ_a}{R} + \frac{y(yp-r)(P-d)FQ_a^2}{2pR} \tag{4.15}$$

Subtracting Equation 4.15 from Equation 4.8, potential cost savings due to a special order,

$$W = dQ_a + \frac{CQ_a}{Q^*} - C + \frac{y(yp-r)Q^*PFQ_a}{2pR} - \frac{yq(P-d)FQ_a}{R} - \frac{y(yp-r)(P-d)FQ_a^2}{2pR} \tag{4.16}$$

In order to maximize the cost savings W, $dW/dQ_a = 0$ shows

$$Q_a^* = \left(d + \frac{C}{Q^*}\right)\frac{pR}{y(yp-r)(P-d)F} + \frac{PQ^*}{2(P-d)} - \frac{qp}{(yp-r)} \tag{4.17}$$

Negative second order derivative can easily be ascertained for optimality. Optimal cost savings W* are obtained by substituting Q_a^* in Equation 4.16.

4.1.1 Neglecting Quality Defects

With reference to just-in-time (JIT) or supply chain environment, strategic relationship may be assumed between the supplier and buyer, and defective items are rejected at the supplier's end. A lot with no defective item reaches the buyer's premises, i.e. y = 1, and optimal results can be stated as follows:

$$Q^* = \sqrt{\frac{2pRC}{(p-r)PF}} \tag{4.18}$$

$$Q_a^* = \left(d + \frac{C}{Q^*}\right)\frac{pR}{(p-r)(P-d)F} + \frac{PQ^*}{2(P-d)} - \frac{qp}{(p-r)} \tag{4.19}$$

$$W = \left(d + \frac{C}{Q^*}\right)Q_a^* - C + \frac{(p-r)Q^*PFQ_a^*}{2pR} - \frac{q(P-d)FQ_a^*}{R} - \frac{(p-r)(P-d)FQ_a^{*2}}{2pR} \tag{4.20}$$

4.1.2 Instantaneous Procurement

For instantaneous procurement, the equations reduce to:

$$Q^* = \sqrt{\frac{2RC}{PF}} \tag{4.21}$$

$$Q_a^* = \left(d + \frac{C}{Q^*}\right)\frac{R}{(P-d)F} + \frac{PQ^*}{2(P-d)} - q \tag{4.22}$$

$$W = \left(d + \frac{C}{Q^*}\right)Q_a^* - C + \frac{Q^* PFQ_a^*}{2R} - \frac{q(P-d)FQ_a^*}{R} - \frac{(P-d)FQ_a^{*2}}{2R} \tag{4.23}$$

Substituting Q* in Equation 4.22,

$$Q_a^* = \frac{dR}{(P-d)F} + \frac{1}{(P-d)}\sqrt{\frac{2PCR}{F}} - q \tag{4.24}$$

$$\text{Or } \frac{(P-d)F}{R}Q_a^* = d + \sqrt{\frac{2CPF}{R}} - \frac{q(P-d)F}{R} \tag{4.25}$$

Substituting Q* in Equation 4.23,

$$W^* = \left(d + \sqrt{\frac{2CPF}{R}} - \frac{q(P-d)F}{R}\right)Q_a^* - C - \frac{(P-d)FQ_a^{*2}}{2R}$$

Substituting Equation 4.25,

$$W^* = \frac{(P-d)FQ_a^{*2}}{2R} - C \tag{4.26}$$

4.2 SENSITIVITY ANALYSIS

Let:

Annual demand R = 1,200 units
Ordering cost C = Rs. 100
Purchase cost per unit, P = Rs. 20
Annual holding cost fraction, F = 0.2
Reduction in the price per unit, d = Rs. 2
Stock position when special order is procured, q = 20 units

Now, from Equation 4.21, $Q^* = \sqrt{\frac{2RC}{PF}}$

$$= \sqrt{\frac{2 \times 1200 \times 100}{20 \times 0.2}} = 244.95$$

From Equation 4.24, optimum special order,

$$Q_a^* = \frac{dR}{(P-d)F} + \frac{1}{(P-d)}\sqrt{\frac{2PCR}{F}} - q$$

$$= \frac{2 \times 1200}{18 \times 0.2} + \frac{1}{18}\sqrt{\frac{2 \times 20 \times 100 \times 1200}{0.2}} - 20$$

$$= 918.83 \text{ units}$$

Optimal cost savings due to a special order using Equation 4.26:

$$W^* = \frac{(P-d)FQ_a^{*2}}{2R} - C$$

$$= \frac{(20-2) \times 0.2 \times 918.83^2}{2 \times 1200} - 100$$

$$= \text{Rs. } 1{,}166.38$$

The following figure represents the situation:

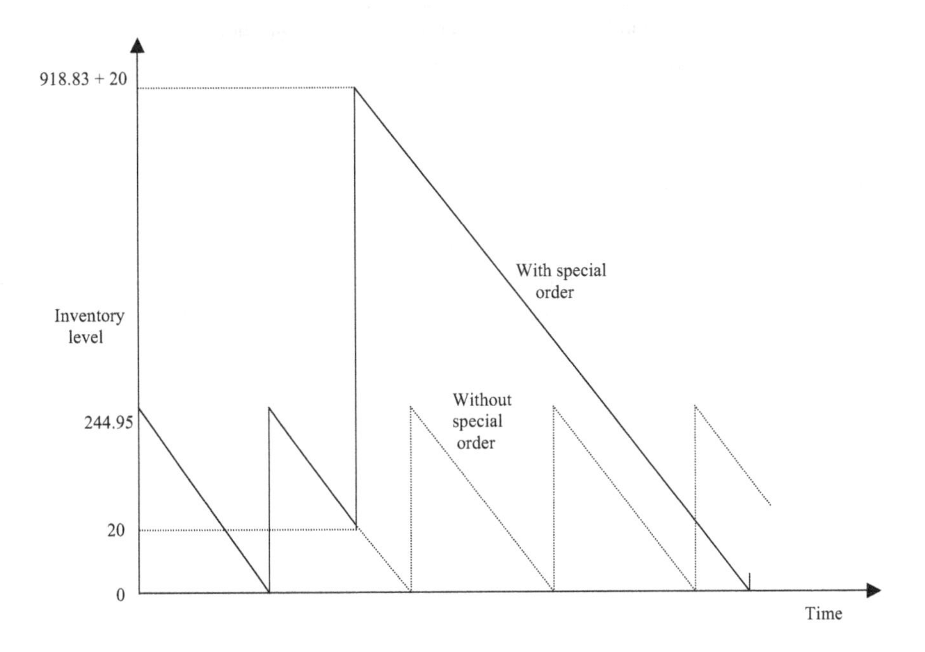

4.2.1 With Respect to Price Discount

For price discount, d = Rs. 2, Q_a* = 918.83 units.

And W* = Rs. 1,166.38

Discount is varied with an increment of Re. 1, and percentage increase in Q_a* and W* are obtained with reference to values 918.83 and Rs. 1,166.38, respectively, corresponding to d = Rs. 2 as follows:

d (Rs.)	Qa* (Units)	% Increase in Qa*	W* (Rs.)	% Increase in W*
3	974.06	6.01	1,244.12	6.66
4	1,036.19	12.77	1,331.58	14.16
5	1,106.60	20.44	1,430.70	22.66
6	1,187.07	29.19	1,543.99	32.37

In order to take the benefit of unit price reduction for the short term, special order quantity Q_a* is procured. Q_a* and optimal cost savings W* are more sensitive towards higher values of temporary price discounts.

4.2.2 With Respect to Annual Holding Cost Fraction

Depending on the overall economic scenario, estimated annual holding cost fraction F varies. It is of interest to conduct sensitivity analysis with respect to F. With the change in F, normal ordering quantity Q* also changes. For different values of F, optimal results are as follows:

F	% Increase in F	Q* (units)	Qa* (units)	W* (Rs.)
0.22	10	233.55	845.56	1,079.70
0.24	20	223.61	784.01	1,006.40
0.26	30	214.83	731.53	943.50
0.28	40	207.02	686.21	888.86
0.30	50	200	646.67	840.90

Consider the original data set in which F = 0.2 and optimal results are Q* = 244.95, Q_a* = 918.83 and W* = Rs. 1,166.38. With reference to these values, the percentage decrease in Q*, Q_a* and W* corresponding to increase in F are evaluated as follows:

% Increase in F	% Decrease in		
	Q*	Qa*	W*
10	4.65	7.97	7.43
20	8.71	14.67	13.72
30	12.29	20.38	19.11
40	15.48	25.32	23.79
50	18.35	29.62	27.90

Sensitivity of the optimal results decreases towards higher values of the annual holding cost fraction. Special order quantity varies more as compared to normal ordering quantity.

4.2.3 With Respect to Ordering Cost

In the JIT or supply chain environment, the emphasis is on the reduction in ordering cost. Corresponding to the reduced ordering cost, C, the results are as follows:

C (Rs.)	Q* (Units)	Qa* (Units)	W* (Rs.)
90	232.38	904.87	1,138.17
80	219.09	890.10	1,108.41
70	204.94	874.38	1,076.80
60	189.74	857.48	1,042.92
50	173.20	839.12	1,006.17

Corresponding to the values of Q*, Q_a* and W* as 244.95, 918.83 and 1,166.38, percentage decrease in the computed values are mentioned in the following table.

% Decrease in C	% Decrease in		
	Q*	Qa*	W*
10	5.13	1.52	2.42
20	10.56	3.13	4.97
30	16.33	4.84	7.68
40	22.54	6.67	10.58
50	29.29	8.67	13.73

The optimal results are more sensitive towards higher decrease in ordering cost. Sensitivity of special order quantity is less as compared to normal ordering quantity. The knowledge of the variation of special ordering quantity beforehand helps in the proper planning of the storage space and operational management.

4.3 PRICE INCREASE

A price increase is often declared well in advance. An organization may place a special order for such items for which price increase in the near future is known. The special order is placed before the price increase becomes effective. This is represented by Figure 4.2.

Let k be the price increase per unit which will be effective at the time t_r onwards. A special order Q_a is placed well before time t_r and the replenishment starts when the stock position is q. The transitional situation between t_r and t_g is under consideration.

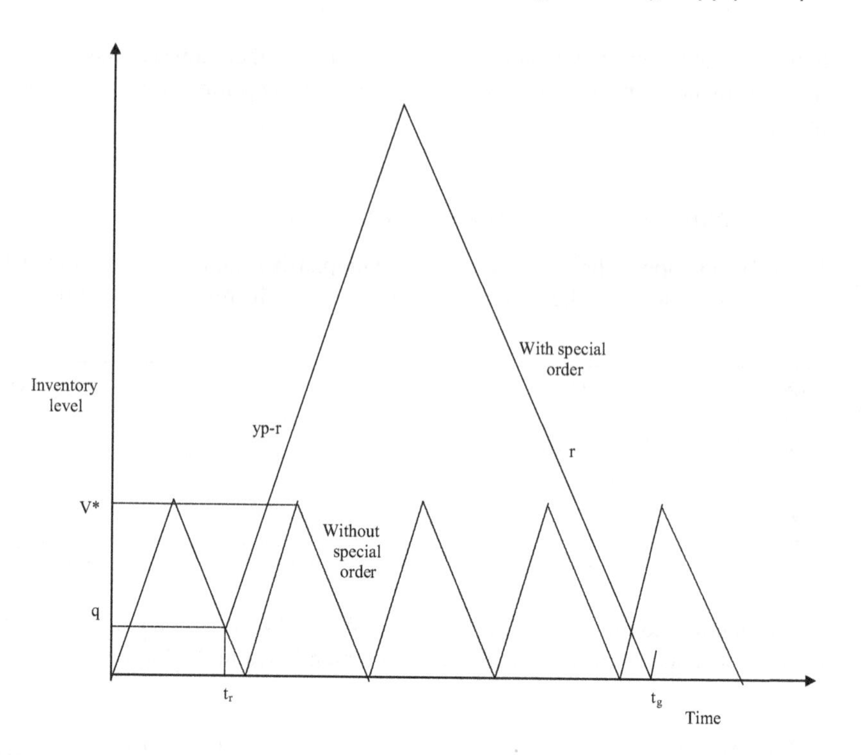

FIGURE 4.2 Price increase situation.

Without a Special Order

Since the price per unit is (P + k) without a special order, Equation 4.1 can be transformed as follows:

$$Q^* = \sqrt{\frac{2pRC}{(yp-r)(P+k)Fy}} \tag{4.27}$$

Total cost was derived in Section 4.1 given by Equation 4.8. P is replaced by (P + k) in Equation 4.8, except in the carrying cost due to stock q which corresponds to price P per unit.

$$\text{Total cost, } T = (P+k)Q_a + \frac{Q_a}{Q^*} \cdot C + \frac{q^2 PF}{2R} + \frac{y(yp-r)Q^* Q_a (P+k)F}{2pR} \tag{4.28}$$

With a Special Order

As the special order Q_a is procured with price P per unit, Equation 4.15 can be written as follows by replacing (P – d) with P:

$$\text{Total Cost, } T_a = C + PQ_a + \frac{q^2 PF}{2R} + \frac{yqPFQ_a}{R} + \frac{y(yp-r)PFQ_a^2}{2pR} \tag{4.29}$$

Subtracting Equation 4.29 from Equation 4.28, potential cost savings due to a special order,

$$W = kQ_a + \frac{C}{Q^*}Q_a - C + \frac{y(yp-r)Q^*(P+k)FQ_a}{2pR} - \frac{yqPFQ_a}{R} - \frac{y(yp-r)PFQ_a^2}{2pR} \tag{4.30}$$

In order to maximize the cost savings W, $dW/dQ_a = 0$ shows

$$Q_a^* = \left(k + \frac{C}{Q^*}\right)\frac{pR}{y(yp-r)PF} + \frac{(P+k)Q^*}{2P} - \frac{qp}{(yp-r)} \tag{4.31}$$

It is easier to ascertain the optimality by confirming the negative second order derivative. By substituting Equation 4.31 in Equation 4.30, optimum cost savings W* are obtained.

4.3.1 Ignoring Defective Items

If y = 1, optimum values are given by:

$$Q^* = \sqrt{\frac{2pRC}{(p-r)(P+k)F}} \tag{4.32}$$

$$Q_a^* = \left(k + \frac{C}{Q^*}\right)\frac{pR}{(p-r)PF} + \frac{(P+k)Q^*}{2P} - \frac{qp}{(p-r)} \tag{4.33}$$

$$W^* = \left(k + \frac{C}{Q^*}\right)Q_a^* - C + \frac{(p-r)Q^*(P+k)FQ_a^*}{2pR} - \frac{qPFQ_a^*}{R} - \frac{(p-r)PFQ_a^{*2}}{2pR} \tag{4.34}$$

4.3.2 Infinite Replenishment Rate

If procurement is instantaneous, the optimal results are given as:

$$Q^* = \sqrt{\frac{2RC}{(P+k)F}} \tag{4.35}$$

$$Q_a^* = \left(k + \frac{C}{Q^*}\right)\frac{R}{PF} + \frac{(P+k)Q^*}{2P} - q \tag{4.36}$$

$$W^* = \left(k + \frac{C}{Q^*}\right)Q_a^* - C + \frac{Q^*(P+k)FQ_a^*}{2R} - \frac{qPFQ_a^*}{R} - \frac{PFQ_a^{*2}}{2R} \tag{4.37}$$

Substituting Equation 4.35 in Equation 4.36,

$$Q_a^* = \frac{kR}{PF} + \frac{1}{P}\sqrt{\frac{2CR(P+k)}{F}} - q \quad (4.38)$$

$$\text{Or } \frac{PFQ_a^*}{R} = k + \sqrt{\frac{2C(P+k)F}{R}} - \frac{qPF}{R} \quad (4.39)$$

Equation 4.37 may be written as follows after putting the value of Q* from Equation 4.35:

$$W^* = \left(k + \sqrt{\frac{2C(P+k)F}{R}} - \frac{qPF}{R}\right)Q_a^* - \frac{PFQ_a^{*2}}{2R} - C$$

Replacing the coefficient of Q_a^* using Equation 4.39,

$$W^* = \frac{PFQ_a^{*2}}{2R} - C \quad (4.40)$$

Example 4.1

Consider:

Annual demand R = 1,200 units
Ordering cost C = Rs. 100
Unit purchase cost P = Rs. 20
Annual carrying cost fraction F = 0.2
Announced increase in price per unit, k = Rs. 2
Position of the stock when special order is procured, q = 20 units.

From Equation 4.35, $Q^* = \sqrt{\frac{2 \times 1200 \times 100}{(20+2) \times 0.2}}$

$= 233.55$ units

From Equation 4.38,

$$Q_a^* = \frac{2 \times 1200}{20 \times 0.2} + \frac{1}{20}\sqrt{\frac{2 \times 100 \times 1200(20+2)}{0.2}} - 20$$

$= 836.90$ units

Optimal cost savings due to a special order using Equation 4.40,

$$W^* = \frac{20 \times 0.2 \times 836.90^2}{2 \times 1200} - 100$$

$= \text{Rs. } 1{,}067.35$

Unit price increase k is varied and computational results are as follows:

k (Rs.)	Q* (Units)	Qa* (Units)	W* (Rs.)
3	228.42	1,142.68	2,076.19
4	223.60	1,448.33	3,396.09
5	219.09	1,753.86	5,026.72
6	214.83	2,059.28	6,967.72

Percentage variation in optimal results are as follows:

k (Rs.)	% Decrease in Q*	% Increase in Qa*
3	2.18	36.53
4	4.23	73.06
5	6.17	109.56
6	7.99	146.06

As compared to price discount situation, in the present case, percentage increase in the optimal special order quantity is very high, along with significant savings—but the arrangement of storage space may become a critical factor in decision making, and the storage space may be a binding constraint.

Example 4.2

In addition to the data R = 1,200 units, C = Rs. 100, P = Rs. 200, F = 0.2, k = Rs. 2 and q = 20 units, assume:

Proportion of nondefective items in a lot, y = 0.95
Replenishment rate per period, p = 120
Demand rate per period, r = 100

The situation refers to the case with finite replenishment rate and Equation 4.27, Equation 4.30 and Equation 4.31 would be applicable.

From Equation 4.27, $Q^* = \sqrt{\dfrac{2pRC}{(yp-r)(P+k)Fy}}$

$$= \sqrt{\frac{2\times 120\times 1200\times 100}{(114-100)\times(200+2)\times 0.2\times 0.95}}$$

$= 231.52$ units

From Equation 4.31, $Q_a^* = \left(k+\dfrac{C}{Q^*}\right)\dfrac{pR}{y(yp-r)PF}+\dfrac{(P+k)Q^*}{2P}-\dfrac{qp}{(yp-r)}$

Substituting the numerical values, $Q_a^* = 603.76$ units. Following the similar procedure as explained in Section 4.3.2, it can be shown that the optimal cost saving due to special order is:

$$W^* = \frac{y(yp-r)PFQ_a^{*2}}{2pR} - C$$

$$= \frac{0.95(114-100)\times 200\times 0.2\times 603.76^2}{2\times 120\times 1200} - 100$$

$$= \text{Rs. } 573.35$$

In a flexible environment, production/replenishment rate may vary. Further with the enhancement in the operational efficiency, there may be an effort to match replenishment rate with demand rate, i.e. yp is slightly higher than r.

From Equation 4.2, optimal maximum inventory during the cycle without special order,

$$V^* = \frac{(yp-r)Q^*}{p}$$

$$\text{At present, } V^* = \frac{(0.95\times 120-100)\times 231.52}{120}$$

$$= 27.01 \text{ units}$$

With the variation in the replenishment rate p, computational results are as follows:

p (Units/Period)	Q* (Units)	V* (Units)	Qa* (Units)	W* (Rs.)
118	246.95	25.32	666.98	622.29
116	266.67	23.45	760.16	704.49
114	293.06	21.34	888.77	810.59

With the decrease in the replenishment rate, optimal ordering quantity increases but the maximum inventory during the normal cycle decreases. The optimal results are more sensitive towards higher decrease in the value of replenishment rate.

4.4 GENERALIZED MODEL

The present section discusses generalized models, which include shortages (along with partial backordering) and finite replenishment rate for the situation:

1. When there is temporary price discount offer
2. When a price increase is declared

4.4.1 Short-Term Price Discount

For a generalized situation including fractional backordering and finite replenishment along with quality level, optimal results are given by Equation 2.46 and

Equation 2.47, discussed in Section 2.2.2. These are as follows, using S as shortage cost:

$$Q^* = \sqrt{\frac{2pRC(S+PF-bP)}{PFy(yp-r)(S-bP)}} \tag{4.41}$$

$$\text{And } J^* = \sqrt{\frac{2RCPF(yp-r)}{yp(S+PF-bP)(S-bP)}} \tag{4.42}$$

Where J* is the optimum shortage quantity and b is a fraction of shortage quantity which is not backordered.

Feasibility condition, i.e. annual shortage cost per unit, S > P, was also discussed,

To deal with the short-term price discount offer, three possibilities may be analyzed:

1. Replenishment with positive stock status, i.e. positive q
2. Replenishment with negative q
3. Replenishment with negative q = J*

All the three possibilities are explored, assuming the withdrawal of inventory items on the first-in-first-out (FIFO) basis.

4.4.1.1 Replenishment with Positive q

This situation is shown in Figure 4.3.

In one normal cycle, shortages exist during the period [J*/(yp – r)] + [J*/r] =[ypJ*/r(yp – r)](r/R) year = ypJ*/R(yp – r) year.

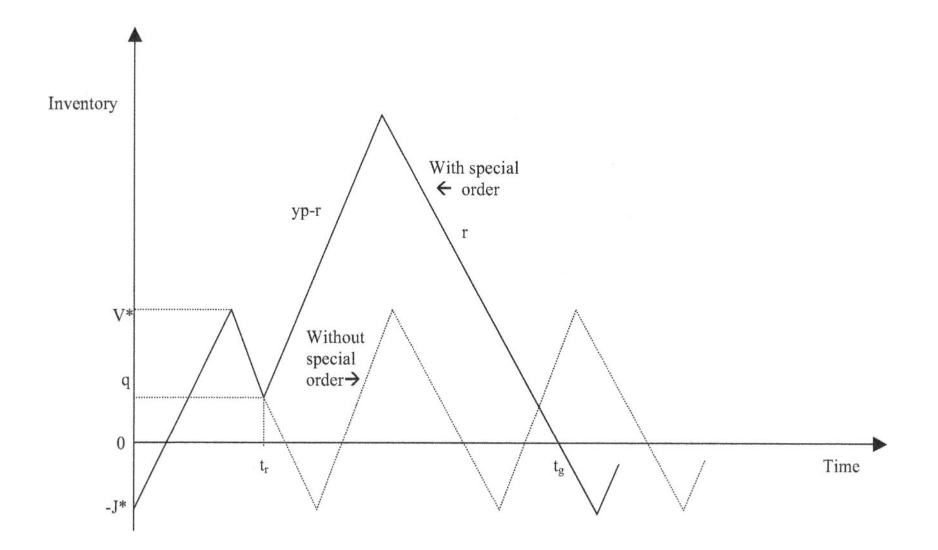

FIGURE 4.3 Transitional price discount situation (positive q).

Shortage quantity during the period $t_r - t_g$ without special order,

$$= \frac{J^*}{2} \cdot \frac{ypJ^*}{R(yp-r)} \cdot \frac{Q_a}{Q^*}$$

$$= \frac{ypJ^{*2} Q_a}{2R(yp-r)Q^*}$$

Shortage quantity which is not backordered $= \dfrac{bypJ^{*2} Q_a}{2R(yp-r)Q^*}$

In one normal cycle, positive inventory exists during (V*/R) + [rV*/(yp – r)R] = ypV*/R(yp – r) year.

Inventory carrying cost during $t_r - t_g$ without special order $= \dfrac{V^*}{2} \cdot \dfrac{ypV^*}{R(yp-r)} \cdot \dfrac{Q_a}{Q^*} \cdot PF$

Total cost without special order during $t_r - t_g$,

$$T = P\left(Q_a - \frac{bypJ^{*2} Q_a}{2R(yp-r)Q^*}\right) - \frac{C}{Q^*}\left(Q_a - \frac{bypJ^{*2} Q_a}{2R(yp-r)Q^*}\right) + \frac{SypJ^{*2} Q_a}{2R(yp-r)Q^*} + \frac{q^2 PF}{2R} + \frac{ypV^{*2} Q_a PF}{2R(yp-r)Q^*} \tag{4.43}$$

Where $V^* = \dfrac{(yp-r)Q^*}{p} - J^*$ (4.44)

and total cost with a special order,

$$T_a = C + (P-d)Q_a + \frac{q^2 PF}{2R} + \frac{yq(P-d)FQ_a}{R} + \frac{y(yp-r)(P-d)FQ_a^2}{2pR} \tag{4.45}$$

Now Equation 4.43 can be written as,

$$T = PQ_a + \frac{CQ_a}{Q^*} + \frac{q^2 PF}{2R} + \frac{Q_a}{Q^*}\left[\frac{(S-bP)ypJ^{*2}}{2R(yp-r)} + \frac{ypV^{*2} PF}{2R(yp-r)}\right] - \frac{CbypJ^{*2} Q_a}{2R(yp-r)Q^{*2}}$$

Substituting V*, J* and Q* from Equation 4.44, Equation 4.42 and Equation 4.41 at appropriate places,

$$T = PQ_a + \frac{CQ_a}{Q^*} + \frac{q^2PF}{2R} - \frac{CbyP^2F^2(yp-r)Q_a}{2pR(S+PF-bP)^2}$$
$$+ \frac{Q_a}{Q^*}\left[\frac{CPF}{(S+PF-bP)} + PF\left\{\frac{C(S+PF-bP)}{PF(S-bP)}\right.\right. \tag{4.46}$$
$$\left.\left. - \frac{2C}{(S-bP)} + \frac{CPF}{(S-bP)(S+PF-bP)}\right\}\right]$$

Now,

$$\frac{CPF}{(S+PF-bP)} + PF\left[\frac{C(S+PF-bP)}{PF(S-bP)} - \frac{2C}{(S-bP)}\right.$$
$$\left. + \frac{CPF}{(S-bP)(S+PF-bP)}\right]$$
$$= \frac{CPF}{(S+PF-bP)} + \frac{PFC}{(S-bP)PF}\left[(S+PF-bP) - 2PF + \frac{P^2F^2}{(S+PF-bP)}\right]$$
$$= \frac{CPF}{(S+PF-bP)} + \frac{C}{(S-bP)}\left[(S-bP) - PF\left\{1 - \frac{PF}{(S+PF-bP)}\right\}\right]$$
$$= \frac{CPF}{(S+PF-bP)} + \frac{C}{(S-bP)}\left[(S-bP) - \frac{PF(S-bP)}{(S+PF-bP)}\right]$$
$$= \frac{CPF}{(S+PF-bP)} + C\left[1 - \frac{PF}{(S+PF-bP)}\right]$$
$$= C$$

Substituting the value in Equation 4.46,

Total cost without special order,

$$T = PQ_a + \frac{CQ_a}{Q^*} + \frac{q^2PF}{2R} - \frac{CbyP^2F^2(yp-r)Q_a}{2pR(S+PF-bP)^2} + \frac{CQ_a}{Q^*}$$

$$\text{Or } T = PQ_a + \frac{2CQ_a}{Q^*} + \frac{q^2PF}{2R} - \frac{CbyP^2F^2(yp-r)Q_a}{2pR(S+PF-bP)^2} \tag{4.47}$$

Subtracting Equation 4.45 from Equation 4.47, potential cost savings due to special order,

$$W = T - T_a$$

$$= Q_a\left[d + \frac{2C}{Q^*} - \frac{yq(P-d)F}{R} - \frac{CbyP^2F^2(yp-r)}{2pR(S+PF-bP)^2}\right]$$

$$-C - \frac{y(yp-r)(P-d)FQ_a^2}{2pR} \tag{4.48}$$

$dW/dQ_a = 0$ shows:

$$Q_a^* = \frac{pR}{y(yp-r)(P-d)F}\left[d + \frac{2C}{Q^*} - \frac{yq(P-d)F}{R} - \frac{CbyP^2F^2(yp-r)}{2pR(S+PF-bP)^2}\right] \tag{4.49}$$

And $W^* = \frac{y(yp-r)(P-d)FQ_a^{*2}}{2pR} - C$ (4.50)

4.4.1.2 Replenishment with Negative q

The situation is shown in Figure 4.4.

Total cost without special order:

$$T = P\left(Q_a - \frac{bypJ^{*2}Q_a}{2R(yp-r)Q^*}\right) + \frac{C}{Q^*}\left(Q_a - \frac{bypJ^{*2}Q_a}{2R(yp-r)Q^*}\right)$$

$$+ \frac{SypJ^{*2}Q_a}{2R(yp-r)Q^*} + \frac{ypV^{*2}Q_aPF}{2R(yp-r)Q^*} + \frac{SJ^{*2}}{2R} - \frac{Sq^2}{2R}$$

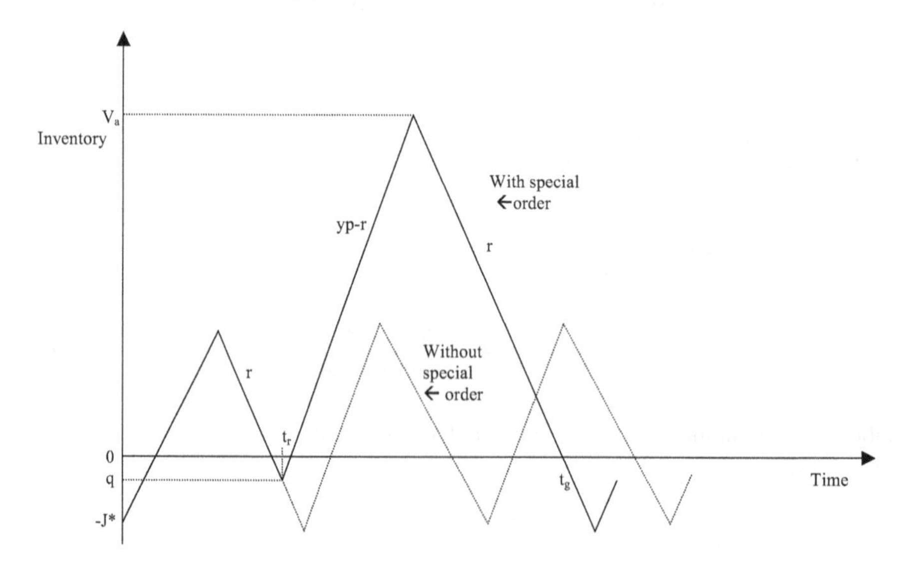

FIGURE 4.4 Transitional price discount situation (negative q).

Following the similar procedure by substituting the optimal values,

$$T = PQ_a + \frac{2CQ_a}{Q^*} + \frac{SJ^{*2}}{2R} - \frac{Sq^2}{2R} - \frac{CbyP^2F^2(yp-r)Q_a}{2pR(S+PF-bP)^2} \tag{4.51}$$

Total cost with a special order,

$$T_a = C + (P-d)Q_a + \frac{yp(P-d)FV_a^2}{2R(yp-r)} + \frac{q^2rS}{2R(yp-r)}$$

$$V_a = \frac{(yp-r)Q_a}{p} - q$$

As ,

$$T_a = C + (P-d)Q_a + \frac{y(yp-r)(P-d)FQ_a^2}{2pR} - \frac{yq(P-d)FQ_a}{R} + \frac{ypF(P-d)q^2}{2R(yp-r)} + \frac{q^2rS}{2R(yp-r)} \tag{4.52}$$

Subtracting Equation 4.52 from Equation 4.51, potential cost savings due to a special order,

$$W = Q_a\left[d + \frac{2C}{Q^*} - \frac{yq(P-d)F}{R} - \frac{CbyP^2F^2(yp-r)}{2pR(S+PF-bP)^2}\right] - C - ypq^2\left[\frac{S+(P-d)F}{2R(yp-r)}\right] + \frac{SJ^{*2}}{2R} - \frac{y(yp-r)(P-d)FQ_a^2}{2pR} \tag{4.53}$$

$dW/dQ_a = 0$ gives:

$$Q_a^* = \frac{pR}{y(yp-r)(P-d)F}\left[d + \frac{2C}{Q^*} + \frac{yq(P-d)F}{R} - \frac{CbyP^2F^2(yp-r)}{2pR(S+PF-bP)^2}\right] \tag{4.54}$$

Comparing this equation with Equation 4.49, it can be said that the previous result is valid for the present case also by using algebraic value of q, as q is negative.

From Equation 4.53,

$$W^* = \frac{y(yp-r)(P-d)FQ_a^{*2}}{2pR} - C + \frac{SJ^{*2}}{2R} - ypq^2\left[\frac{S+(P-d)F}{2R(yp-r)}\right] \tag{4.55}$$

4.4.1.3 Replenishment with Negative q = J*

This is shown in Figure 4.5.

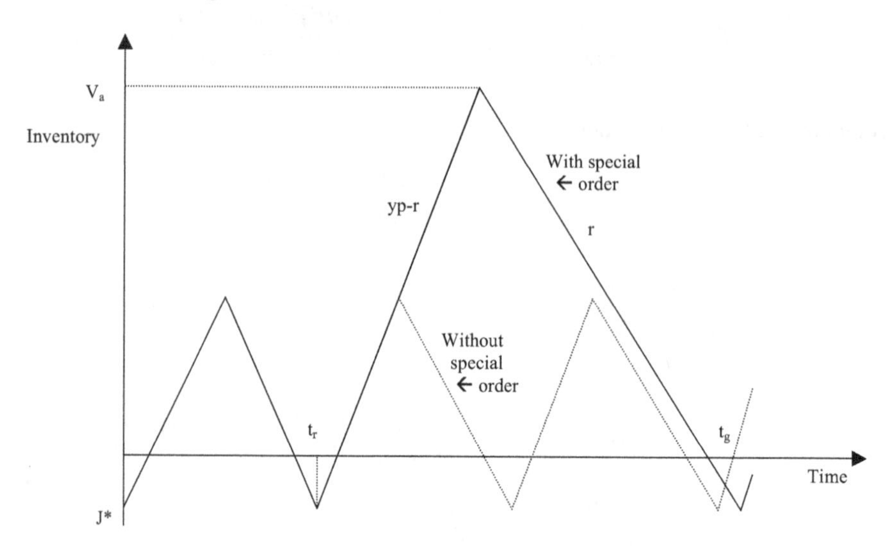

FIGURE 4.5 Transitional price discount situation (negative q = J*).

For the case without special order, one order is placed for the unit price (P – d) and after that, the unit price applicable is P.

Total cost without special order,

$$T = (P-d)Q^* + P\left(Q_a - Q^* - \frac{bypJ^{*2}Q_a}{2R(yp-r)Q^*}\right)$$
$$+\frac{C}{Q^*}\left(Q_a - \frac{bypJ^{*2}Q_a}{2R(yp-r)Q^*}\right) + \frac{SypJ^{*2}Q_a}{2R(yp-r)Q^*}$$
$$+\frac{ypV^{*2}(P-d)F}{2R(yp-r)} + \frac{ypV^{*2}PF(Q_a - Q^*)}{2R(yp-r)}$$

$$= PQ_a + \frac{CQ_a}{Q^*} - dQ^* + \frac{Q_a}{Q^*}\left[\frac{(S-bP)ypJ^{*2}}{2R(yp-r)} + \frac{ypV^{*2}PF}{2R(yp-r)}\right]$$
$$-\frac{CbypJ^{*2}Q_a}{2R(yp-r)Q^{*2}} - \frac{ypV^{*2}dF}{2R(yp-r)}$$

Substituting the values,

$$T = PQ_a + \frac{2CQ_a}{Q^*} - \frac{CbyP^2F^2(yp-r)Q_a}{2pR(S+PF-bP)^2} - dQ^* - \frac{dC(S-bP)}{P(S+PF-bP)} \tag{4.56}$$

Total cost with a special order,

$$T_a = C + (P-d)Q_a + \frac{yp(P-d)FV_a^2}{2R(yp-r)} + \frac{J^{*2}rS}{2R(yp-r)}$$

As $V_a = \frac{(yp-r)Q_a}{p} - J^*$,

$$T_a = C + (P-d)Q_a + \frac{y(yp-r)(P-d)FQ_a^2}{2pR} - \frac{yJ^*(P-d)FQ_a}{R} + \frac{ypF(P-d)J^{*2}}{2R(yp-r)} + \frac{J^{*2}rS}{2R(yp-r)} \quad (4.57)$$

Subtracting Equation 4.57 from Equation 4.56,

$$W = Q_a\left[d + \frac{2C}{Q^*} + \frac{yJ^*(P-d)F}{R} - \frac{CbyP^2F^2(yp-r)}{2pR(S+PF-bP)^2}\right] - \frac{y(yp-r)(P-d)FQ_a^2}{2pR} - C - dQ^* - \frac{dC(S-bP)}{P(S+PF-bP)} - \frac{ypF(P-d)J^{*2}}{2R(yp-r)} - \frac{SrJ^{*2}}{2R(yp-r)} \quad (4.58)$$

$dW/dQ_a = 0$ shows:

$$Q_a^* = \frac{pR}{y(yp-r)(P-d)F}\left[d + \frac{2C}{Q^*} + \frac{yJ^*(P-d)F}{R} - \frac{CbyP^2F^2(yp-r)}{2pR(S+PF-bP)^2}\right] \quad (4.59)$$

Again comparing with Equation 4.49, it can be said that the above equation i.e. Equation 4.59 is obtained by substituting q = –J* in Equation 4.49. Equation 4.59 is also similar to Equation 4.54, except q is replaced by J*.

Now substituting Q_a* in Equation 4.58,

$$W^* = \frac{y(yp-r)(P-d)FQ_a^{*2}}{2pR} - C - dQ^* - \frac{dC(S-bP)}{P(S+PF-bP)} - \frac{ypF(P-d)J^{*2}}{2R(yp-r)} - \frac{SrJ^{*2}}{2R(yp-r)} \quad (4.60)$$

4.4.2 Declared Price Increase

The special order is placed well before the price increase becomes effective. It may be replenished at various stock statuses.

4.4.2.1 Replenishment with Positive q

Since the price per unit is (P + k) without special order,

$$Q^* = \sqrt{\frac{2pRC\left[S+(P+k)F-(P+k)b\right]}{(P+k)Fy(yp-r)\left[S-b(P+k)\right]}} \tag{4.61}$$

And $$J^* = \sqrt{\frac{2RC(P+k)F(yp-r)}{yp\left[S+(P+k)F-b(P+k)\right]\left[S-b(P+k)\right]}} \tag{4.62}$$

Total cost without special order,

$$T=(P+k)\left(Q_a-\frac{bypJ^{*2}Q_a}{2R(yp-r)Q^*}\right)-\frac{C}{Q^*}\left(Q_a-\frac{bypJ^{*2}Q_a}{2R(yp-r)Q^*}\right)$$
$$+\frac{SypJ^{*2}Q_a}{2R(yp-r)Q^*}+\frac{q^2PF}{2R}+\frac{ypV^{*2}Q_a(P+k)F}{2R(yp-r)Q^*}$$

Or $$T=(P+k)Q_a+\frac{2CQ_a}{Q^*}+\frac{q^2PF}{2R}-\frac{Cby(P+k)^2F^2(yp-r)Q_a}{2pR\left[S+(P+k)F-b(P+k)\right]^2} \tag{4.63}$$

Total cost with a special order,

$$T_a=C+PQ_a+\frac{q^2PF}{2R}+\frac{yqPFQ_a}{R}+\frac{y(yp-r)PFQ_a^2}{2pR} \tag{4.64}$$

Now potential cost saving,

$$W=T-T_a$$
$$=Q_a\left[k+\frac{2C}{Q^*}-\frac{yqPF}{R}-\frac{Cby(P+k)^2F^2(yp-r)}{2pR\{S+(P+k)F-b(P+k)\}^2}\right]$$
$$-C-\frac{y(yp-r)PFQ_a^2}{2pR} \tag{4.65}$$

Following similar procedure,

$$Q_a^*=\frac{pR}{y(yp-r)(P-d)F}\left[k+\frac{2C}{Q^*}-\frac{yqPF}{R}\right.$$
$$\left.-\frac{Cby(P+k)^2F^2(yp-r)}{2pR\{S+(P+k)F-b(P+k)\}^2}\right] \tag{4.66}$$

And $$W^*=\frac{y(yp-r)PFQ_a{}^{*2}}{2pR}-C \tag{4.67}$$

4.4.2.2 Replenishment with Negative q

Now, without special order,

$$T = (P+k)\left(Q_a - \frac{bypJ^{*2}Q_a}{2R(yp-r)Q^*}\right) + \frac{C}{Q^*}\left(Q_a - \frac{bypJ^{*2}Q_a}{2R(yp-r)Q^*}\right)$$
$$+\frac{SypJ^{*2}Q_a}{2R(yp-r)Q^*} + \frac{ypV^{*2}Q_a(P+k)F}{2R(yp-r)Q^*} + \frac{SJ^{*2}}{2R} - \frac{Sq^2}{2R}$$

Or

$$T = (P+k)Q_a + \frac{2CQ_a}{Q^*} + \frac{SJ^{*2}}{2R} - \frac{Sq^2}{2R}$$
$$-\frac{Cby(P+k)^2F^2(yp-r)Q_a}{2pR\left[S+(P+k)F-b(P+k)\right]^2} \tag{4.68}$$

And with special order,

$$T_a = C + PQ_a + \frac{y(yp-r)PFQ_a^2}{2pR} - \frac{yqPFQ_a}{R} + \frac{ypFPq^2}{2R(yp-r)}$$
$$+\frac{q^2rS}{2R(yp-r)} \tag{4.69}$$

And,

$$W = Q_a\left[k + \frac{2C}{Q^*} - \frac{yqPF}{R} - \frac{Cby(P+k)^2F^2(yp-r)}{2pR\{S+(P+k)F-b(P+k)\}^2}\right]$$
$$-C - ypq^2\left[\frac{S+PF}{2R(yp-r)}\right] + \frac{SJ^{*2}}{2R} - \frac{y(yp-r)PFQ_a^2}{2pR} \tag{4.70}$$

$dW/dQ_a = 0$ shows,

$$Q_a^* = \frac{pR}{y(yp-r)PF}\left[k + \frac{2C}{Q^*} + \frac{yqPF}{R}\right.$$
$$\left.-\frac{Cby(P+k)^2F^2(yp-r)}{2pR\{S+(P+k)F-b(P+k)\}^2}\right] \tag{4.71}$$

and $$W^* = \frac{y(yp-r)PFQ_a^{*2}}{2pR} - C + \frac{SJ^{*2}}{2R} - ypq^2\left[\frac{S+PF}{2R(yp-r)}\right] \tag{4.72}$$

4.4.2.3 Replenishment with Negative q = J*

For the case without special order, a single order is placed for the unit price P, and after that, unit price applicable is (P + k).

$$T = PQ^* + (P+k)\left(Q_a - Q^* - \frac{bypJ^{*2}Q_a}{2R(yp-r)Q^*}\right)$$
$$+\frac{C}{Q^*}\left(Q_a - \frac{bypJ^{*2}Q_a}{2R(yp-r)Q^*}\right) + \frac{SypJ^{*2}Q_a}{2R(yp-r)Q^*}$$
$$+\frac{ypV^{*2}PF}{2R(yp-r)} + \frac{ypV^{*2}(P+k)F(Q_a - Q^*)}{2R(yp-r)}$$

$$= (P+k)Q_a - kQ^* + \frac{q^2PF}{2R} + \frac{Q_a}{Q^*}\left[\frac{\{S-b(P+k)\}ypJ^{*2}}{2R(yp-r)}\right.$$
$$\left.+\frac{ypV^{*2}(P+k)F}{2R(yp-r)}\right] - \frac{CbypJ^{*2}Q_a}{2R(yp-r)Q^{*2}} - \frac{ypV^{*2}kF}{2R(yp-r)}$$

Or

$$T = (P+k)Q_a + \frac{2CQ_a}{Q^*} - \frac{Cby(P+k)^2F^2(yp-r)Q_a}{2pR\left[S+(P+k)F-b(P+k)\right]^2}$$
$$-kQ^* - \frac{kC\left[S-b(P+k)\right]}{(P+k)\left[S+(P+k)F-b(P+k)\right]} \tag{4.73}$$

With special order,

$$T_a = C + PQ_a + \frac{y(yp-r)PFQ_a^2}{2pR} - \frac{yJ^*PFQ_a}{R} + \frac{ypFPJ^{*2}}{2R(yp-r)}$$
$$+\frac{J^{*2}rS}{2R(yp-r)} \tag{4.74}$$

and,

$$W = Q_a\left[k + \frac{2C}{Q^*} + \frac{yJ^*PF}{R} - \frac{Cby(P+k)^2F^2(yp-r)}{2pR\{S+(P+k)F-b(P+k)\}^2}\right]$$
$$-\frac{y(yp-r)PFQ_a^2}{2pR} - C - kQ^* - \frac{kC\left[S-b(P+k)\right]}{(P+k)\left[S+(P+k)F-b(P+k)\right]}$$
$$-\frac{ypFPJ^{*2}}{2R(yp-r)} - \frac{SrJ^{*2}}{2R(yp-r)} \tag{4.75}$$

Following the similar procedure,

$$Q_a^* = \frac{pR}{y(yp-r)PF}\left[k + \frac{2C}{Q^*} + \frac{yJ^*PF}{R}\right.$$
$$\left.-\frac{Cby(P+k)^2F^2(yp-r)}{2pR\{S+(P+k)F-b(P+k)\}^2}\right] \tag{4.76}$$

And

$$W^* = \frac{y(yp-r)PFQ_a^{*2}}{2pR} - C - kQ^* - \frac{kC\left[S-b(P+k)\right]}{(P+k)\left[S+(P+k)F-b(P+k)\right]} - \frac{ypFPJ^{*2}}{2R(yp-r)} - \frac{SrJ^{*2}}{2R(yp-r)} \tag{4.77}$$

The optimal results are obtained in the generalized model for all the three possibilities. In the situation for price increase also, algebraic value of q may be placed in Equation 4.66 which refers to the special order quantity formulation corresponding to replenishment with positive q. Depending on the case, q may be positive, negative or equal to $-J^*$.

When all the shortages are backordered, i.e. b is equal to zero, then the optimal results for temporary price discounts, as well as price increase, are applicable by putting $b = 0$.

The generalized models are not only suitable for replenishment/procurement, but also for the production situation, particularly the price increase case. The production cost may go high due to various reasons such as increase in input item cost and electricity charges, revised salaries, etc. A special manufacturing quantity may be planned before a cost increase is effective in order to take the benefit of potential cost savings.

Exercises

1. What do you understand by the following:
   a. Temporary price discount
   b. Price increase
2. What are the potential benefits of the temporary price discount offers to:
   a. Suppliers
   b. Buyers
3. Discuss the effects of a declared price increase on:
   a. Suppliers
   b. Buyers
4. What type of factors should be considered by the decision maker before responding to:
   a. A declared price increase
   b. A short-term price discount offer
5. Under what situations might a decision to purchase large quantities of a component/input item during the price discount offer period go wrong?
6. Under what situations might a decision to purchase large quantities of a component/input item well before the price increase would be effective go wrong?

7. The following figure represents the case of instantaneous procurement:

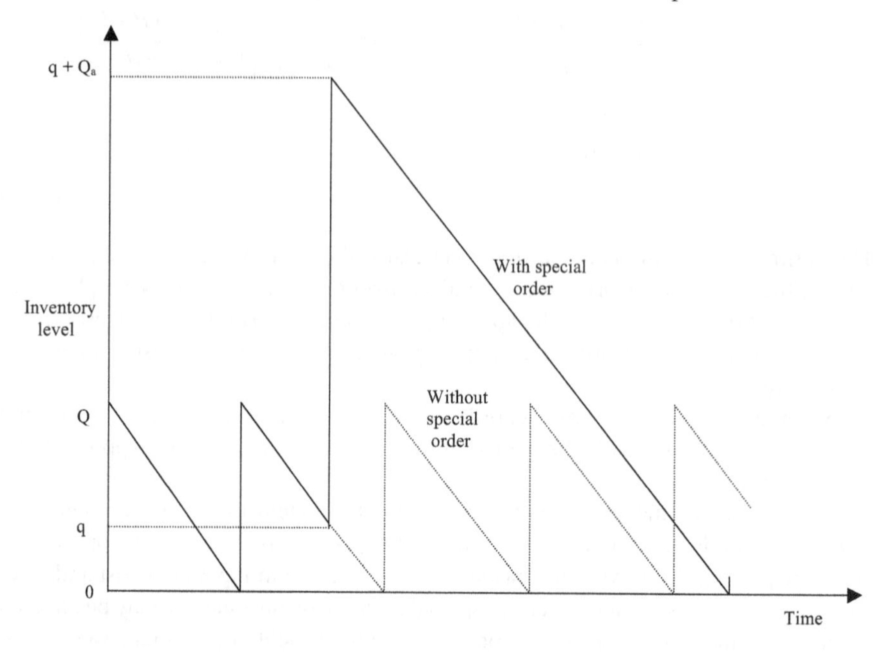

Obtain the optimum expressions for price discount situation concerning:

a. Special order quantity
b. Potential cost savings due to a special order

8. Let:

Annual demand = 1,200 units
Ordering cost = Rs. 120
Purchase cost per unit = Rs. 15
Annual holding cost fraction = 0.25
Reduction in the price per unit = Rs. 3
Stock position when special order is procured = 25 units

a. Find out:
   i. Optimum order quantity
   ii. Special order quantity
   iii. Potential cost savings due to a special order
b. Carry out the sensitivity analysis with respect to:
   i. Price discount
   ii. Annual holding cost fraction
   iii. Ordering cost

9. Include the shortages which are completely backlogged in Exercise 7 and obtain the optimum expressions for:

    a. Special order quantity
    b. Potential cost savings due to a special order

10. Assume that all the shortages are completely backlogged. Additionally:

    Annual backlogging cost per unit = Rs. 150
    Annual demand = 1,200 units
    Stock position when special order is procured = 15 units
    Purchase cost per unit = Rs. 12
    Annual holding cost fraction = 0.26
    Ordering cost = Rs. 115
    Reduction in the price per unit = Rs. 3

    a. Find out:

        i. Optimum ordering quantity
        ii. Maximum shortage quantity
        iii. Special order quantity
        iv. Potential cost savings due to a special order

    b. Carry out the sensitivity analysis with respect to:

        i. Price discount
        ii. Annual holding cost fraction
        iii. Ordering cost
        iv. Annual backlogging cost

11. Include the shortages which are partially backlogged in Exercise 7 and obtain the optimum expressions for:

    a. Special order quantity
    b. Potential cost savings due to a special order

12. Assume that all the shortages are not completely backlogged. Additionally:

    Fraction of shortage quantity which is not backlogged = 0.25
    Annual backlogging cost per unit = Rs. 130
    Annual demand = 1,200 units
    Stock position when special order is procured = 13 units
    Purchase cost per unit = Rs. 14
    Annual holding cost fraction = 0.27
    Ordering cost = Rs. 120
    Reduction in the price per unit = Rs. 2

    a. Evaluate:
        i. Optimum ordering quantity
        ii. Maximum shortage quantity

iii. Special order quantity
iv. Potential cost savings due to a special order

b. Carry out the sensitivity analysis with respect to:

i. Price discount
ii. Annual holding cost fraction
iii. Ordering cost
iv. Annual backlogging cost
v. Fraction of shortage quantity which is not backlogged

13. The following figure represents the case of finite replenishment rate. In order to take the benefit of price discount offer by the supplying firm, the company wants to place an order of larger quantities before the offer ends. Replenishment of special order starts when a certain positive stock of that item still exists. A general approach to the problem would be to compare the situation without a special order and with a special order.

Obtain the optimum generalized expressions for:

a. Special order quantity
b. Potential cost savings due to a special order

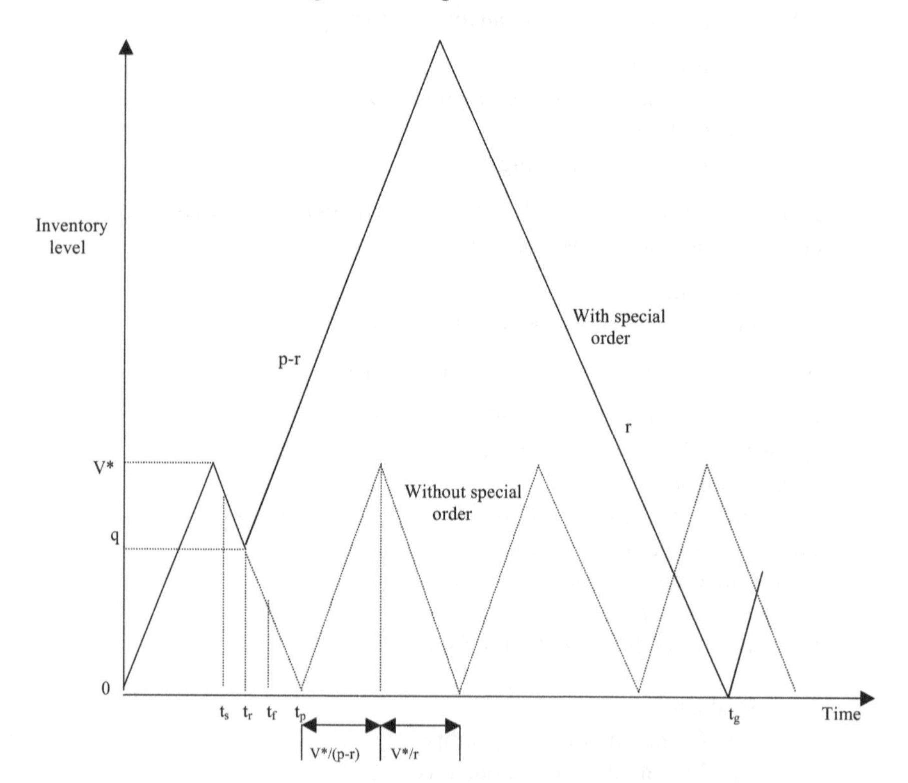

14. Let:

    Demand rate per period = 100
    Replenishment rate per period = 125
    Annual demand = 1,200 units
    Ordering cost = Rs. 120
    Purchase cost per unit = Rs. 15
    Annual holding cost fraction = 0.25
    Reduction in the price per unit = Rs. 3
    Stock position when special order is procured = 25 units

    a. Evaluate:
        i. Optimum order quantity
        ii. Special order quantity
        iii. Potential cost savings due to a special order
    b. Carry out the sensitivity analysis with respect to:
        i. Price discount
        ii. Annual holding cost fraction
        iii. Ordering cost

15. Include the shortages which are completely backlogged in Exercise 13 and obtain the optimum expressions for:
    a. Special order quantity
    b. Potential cost savings due to a special order

16. Assume that all the shortages are completely backordered. Additionally:

    Annual backordering cost per unit = Rs. 150
    Annual demand = 1,200 units
    Demand rate per period = 100
    Replenishment rate per period = 125
    Stock position when special order is procured = 15 units
    Purchase cost per unit = Rs. 12
    Annual holding cost fraction = 0.26
    Ordering cost = Rs. 115
    Reduction in the price per unit = Rs. 3

    a. Find out:
        i. Optimum ordering quantity
        ii. Maximum shortage quantity
        iii. Special order quantity
        iv. Potential cost savings due to a special order
    b. Carry out the sensitivity analysis with respect to:
        i. Price discount
        ii. Annual holding cost fraction

iii. Ordering cost
iv. Annual backlogging cost

17. Include the shortages which are fractionally backlogged in the Exercise 13 and obtain the optimum expressions for:

    a. Special order quantity
    b. Potential cost savings due to a special order

18. Assume that all the shortages are fractionally backlogged. Additionally:

    Fraction of shortage quantity which is not backlogged = 0.25
    Annual backlogging cost per unit = Rs. 130
    Annual demand = 1,200 units
    Stock position when special order is procured = 13 units
    Purchase cost per unit = Rs. 14
    Annual holding cost fraction = 0.27
    Ordering cost = Rs. 120
    Demand rate per period = 100
    Replenishment rate per period = 125
    Reduction in the price per unit = Rs. 2

    a. Calculate:

        i. Optimum ordering quantity
        ii. Maximum shortage quantity
        iii. Special order quantity
        iv. Potential cost savings due to a special order

    b. Carry out the sensitivity analysis with respect to:

        i. Price discount
        ii. Annual holding cost fraction
        iii. Ordering cost
        iv. Annual backlogging cost
        v. Fraction of shortage quantity which is not backlogged

19. Consider the figure included in Exercise 7. Obtain the optimum expressions for declared price increase situation concerning:

    a. Special order quantity
    b. Potential cost savings due to a special order

20. Let:

    Annual demand = 1,200 units
    Ordering cost = Rs. 120
    Purchase cost per unit = Rs. 15
    Annual holding cost fraction = 0.25
    Announced increase in the price per unit = Rs. 3
    Stock position when special order is procured = 25 units

a. Find out:

i. Optimum order quantity
ii. Special order quantity
iii. Potential cost savings due to a special order

b. Carry out the sensitivity analysis with respect to:

i. Price increase
ii. Annual holding cost fraction
iii. Ordering cost

21. Include the shortages which are completely backlogged in Exercise 19 and obtain the optimum expressions related to declared price increase scenario concerning:

a. Special order quantity
b. Potential cost savings due to a special order

22. Assume that all the shortages are completely backlogged. Additionally:

Annual backlogging cost per unit = Rs. 150
Annual demand = 1,200 units
Stock position when special order is procured = 15 units
Purchase cost per unit = Rs. 12
Annual holding cost fraction = 0.26
Ordering cost = Rs. 115
Increase in the price per unit = Rs. 3

a. Find out:

i. Optimum ordering quantity
ii. Maximum shortage quantity
iii. Special order quantity
iv. Potential cost savings due to a special order

b. Carry out the sensitivity analysis with respect to:

i. Price increase
ii. Annual holding cost fraction
iii. Ordering cost
iv. Annual backlogging cost

23. Include the shortages which are partially backlogged in Exercise 19 and obtain the optimum expressions related to declared price increase scenario concerning:

a. Special order quantity
b. Potential cost savings due to a special order

24. Assume that all the shortages are not completely backlogged. Additionally:

Fraction of shortage quantity which is not backlogged = 0.25
Annual backlogging cost per unit = Rs. 130

Annual demand = 1,200 units
Stock position when special order is procured = 13 units
Purchase cost per unit = Rs. 14
Annual holding cost fraction = 0.27
Ordering cost = Rs. 120
Increase in the price per unit = Rs. 2

a. Evaluate:
    i. Optimum ordering quantity
    ii. Maximum shortage quantity
    iii. Special order quantity
    iv. Potential cost savings due to a special order
b. Carry out the sensitivity analysis with respect to:
    i. Price increase
    ii. Annual holding cost fraction
    iii. Ordering cost
    iv. Annual backlogging cost
    v. Fraction of shortage quantity which is not backlogged

25. Consider Exercise 19. Model the finite replenishment rate in the announced price increase situation and derive the optimum values of:
    a. Special order quantity
    b. Potential cost savings due to a special order
26. Let:

    Demand rate per period = 100
    Replenishment rate per period = 125
    Annual demand = 1,200 units
    Ordering cost = Rs. 120
    Purchase cost per unit = Rs. 15
    Annual holding cost fraction = 0.25
    Increase in the price per unit = Rs. 3
    Stock position when special order is procured = 25 units

    a. Evaluate:
        i. Optimum order quantity
        ii. Special order quantity
        iii. Potential cost savings due to a special order
    b. Carry out the sensitivity analysis with respect to:
        i. Price increase
        ii. Annual holding cost fraction
        iii. Ordering cost
27. Include the shortages which are completely backlogged in Exercise 25 and obtain the optimum expressions for:

a. Special order quantity
b. Potential cost savings due to a special order

28. Assume that all the shortages are completely backordered. Additionally:

    Annual backordering cost per unit = Rs. 150
    Annual demand = 1,200 units
    Demand rate per period = 100
    Replenishment rate per period = 125
    Stock position when special order is procured = 15 units
    Purchase cost per unit = Rs. 12
    Annual holding cost fraction = 0.26
    Ordering cost = Rs. 115
    Increase in the price per unit = Rs. 3

    a. Find out:
        i. Optimum ordering quantity
        ii. Maximum shortage quantity
        iii. Special order quantity
        iv. Potential cost savings due to a special order

    b. Carry out the sensitivity analysis with respect to:
        i. Price increase
        ii. Annual holding cost fraction
        iii. Ordering cost
        iv. Annual backlogging cost

29. Include the shortages which are fractionally backlogged in Exercise 25 and obtain the optimum expressions for:

    a. Special order quantity
    b. Potential cost savings due to a special order

30. Assume that all the shortages are fractionally backlogged. Additionally:

    Fraction of shortage quantity which is not backlogged = 0.25
    Annual backlogging cost per unit = Rs. 130
    Annual demand = 1,200 units
    Stock position when special order is procured = 13 units
    Purchase cost per unit = Rs. 14
    Annual holding cost fraction = 0.27
    Ordering cost = Rs. 120
    Demand rate per period = 100
    Replenishment rate per period = 125
    Increase in the price per unit = Rs. 2

    a. Calculate:
        i. Optimum ordering quantity
        ii. Maximum shortage quantity

iii. Special order quantity
iv. Potential cost savings due to a special order

b. Carry out the sensitivity analysis with respect to:

i. Price increase
ii. Annual holding cost fraction
iii. Ordering cost
iv. Annual backlogging cost
v. Fraction of shortage quantity which is not backlogged

31. Explain the generalized models with the help of appropriate figures (including quality defects, partial backordering, and finite replenishment) related to temporary price discounts for replenishment with variety of stock positions such as:

a. Positive stock status
b. Negative stock status
c. Negative stock equivalent to optimal maximum shortage quantity

Also derive the optimum expression for:

a. Special order quantity
b. Potential cost savings due to a special order

32. Explain the generalized models with the help of appropriate figures (including quality defects, partial backordering, and finite replenishment) related to declared price increase for replenishment with variety of stock positions such as:

a. Positive stock status
b. Negative stock status
c. Negative stock equivalent to optimal maximum shortage quantity

Also derive the optimum expression for:

a. Special order quantity
b. Potential cost savings due to a special order

5 Batch Size for Material Requirements Planning

A statement showing how much quantity of an end item is to be produced, and in which time period, is called the master production schedule (MPS). Based on the MPS, the material requirements planning (MRP) system is used to plan for the need of components, input items, etc., so that the items are available at their times of need. The planning of production or procurement orders for items becomes convenient by using MRP. Consider the following net requirements or demands in respective periods:

Period:	**1**	**2**	**3**	**4**	**5**	**6**	**7**	**8**	**9**
Net Demand:	50	45	47	52	51	11	22	13	35

Period may be in days, months or weeks (usually, i.e. weeks No. 1, 2, 3, . . . are mentioned). In the first period or first week, net demand is of 50 items. Similarly, in weeks 2–3, demand is for 45 items and 47 items, respectively, and so on. It is not necessary to procure or produce the quantity as per the demand in that period itself. Clubbing of demands may also take place. For instance, it is the policy of the company to procure for the demands of three periods, and the procurement schedule is as follows:

Period:	**1**	**2**	**3**	**4**	**5**	**6**	**7**	**8**	**9**
Demand:	50	45	47	52	51	11	22	13	35
Procurement Schedule:	142	0	0	114	0	0	70	0	0

Procurement in the Period 1 is for the sum of three period demands, i.e., 50 + 45 + 47 = 142 units. No procurement is there in Period 2 or 3, and therefore zero quantity is mentioned corresponding to these periods. Similarly in Period 4, a total quantity of 114 units (52 + 51 + 11) units are scheduled for purchase and in Period 7, an order of 70 units is placed. In this way, three orders are placed for quantities of 142, 114, and 70 units.

MRP lot sizing refers to the determination of lot sizes in this environment either for production or for procurement. Various MRP lot sizing techniques are discussed further.

5.1 MATERIAL REQUIREMENTS PLANNING (MRP) LOT SIZING TECHNIQUES

Total cost is the significant parameter to be evaluated while obtaining lot sizes using different techniques. Total cost usually consists of ordering/setup costs and inventory carrying costs. This is illustrated considering the example discussed before.

DOI: 10.1201/9781003213994-5

Period:	1	2	3	4	5	6	7	8	9
Demand:	50	45	47	52	51	11	22	13	35
Procurement Schedule:	142	0	0	114	0	0	70	0	0
Balance:	92	47	0	62	11	0	48	35	0

In the first period, a lot of 142 is procured and available balance at the end of Period 1 is 92 (142 – 50). In Period 2, nothing is procured and therefore available balance is 47 (92 + 0 – 45). Similarly in Period 3, balance is zero (47 + 0 – 47).

For Period 4, previous balance (of Period 3) is zero and therefore balance inventory is 62 (0 + 114 – 52) units. Similar procedures are carried out for the remaining periods.

Let, fixed ordering cost per order, C = Rs. 150.

As three orders are placed in periods 1, 4 and 7, total ordering cost = 3 × 150 = Rs. 450.

If F is the inventory carrying cost per unit per period and F = Rs. 2, then inventory carrying cost in period 1 = 92 × 2 = Rs. 184 since balance inventory at the end of Period 1 is 92.

Similarly, carrying cost for Period 2 = 47 × 2 = Rs. 94, and for Period 3, carrying cost is zero because balance inventory is zero in that period.

Total inventory carrying cost = (92 × 2) + (47 × 2) + (0 × 2)
+ (62 × 2) + (11 × 2) + (0 × 2) + (48 × 2) + (35 × 2) + (0 × 2)
= 2 × [sum of balance inventory]
= 2 × [92+47 + 0+62+ 11 +0+48 + 35 +0] = 2 × 295 = Rs. 590.

As the total cost consists of ordering and inventory carrying cost, total cost

= 450 + 590 = Rs. 1,040.

Different MRP lot sizing techniques are explained in the following sections.

5.1.1 Lot for Lot (LFL)

A lot of quantity equal to the net demand in a period is procured or produced in the same period.

For example:

Period:	1	2	3	4	5	6	7	8
Net demand:	20	26	0	11	17	0	59	13
Production/ Procurement Schedule:	20	26	0	11	17	0	59	13
Balance:	0	0	0	0	0	0	0	0

Demand for Period 2 is 26, and 26 units of the item are produced in Period 2, and similarly for the remaining periods. Inventory carrying cost is zero, as the balance is not there in each period. Six orders are placed because in two periods, viz. periods 3 and 6, no order was placed since no requirement was there. If setup/ordering cost is Rs. 50, then total ordering cost = No. of orders × 50

$$= 6 \times 50 = \text{Rs. } 300$$

As the carrying cost is zero using LFL technique, total cost is Rs. 300. In the situation of high carrying cost per unit per period and low ordering cost per order, this method is very useful.

5.1.2 Least Total Cost (LTC)

In this method, the demand for successive periods are clubbed one after another and inventory carrying cost is compared with ordering cost. Stopping criteria is the period for which holding cost is greater than ordering cost, and the demands prior to that period are accumulated as lot size. Consider:

Period:	**1**	**2**	**3**	**4**	**5**	**6**	**7**	**8**
Demand:	20	26	0	11	17	0	59	13

Let ordering cost = Rs. 100 and carrying cost = Rs. 2/unit/period. Start from Period 1. If 20 units are scheduled in that period, then ordering cost of Rs. 100 plus zero inventory cost—as balance inventory is zero in that period—form the total cost as Rs. 100, i.e.:

Period:	**1**
Demand:	20
Lot Size:	20
Balance:	0

Now, if the demand of Period 1 and Period 2 are clubbed together, then:

Period:	**1**	**2**
Demand:	20	26
Lot Size:	46	0
Balance:	26	0

Ordering cost = Rs. 100
Inventory carrying cost = 26 × 2 = Rs. 52

Now Period 3 demand, which is zero, is included:

Period:	**1**	**2**	**3**
Demand:	20	26	0

Period:	1	2	3
Lot Size:	46	0	0
Balance:	26	0	0

This gives the ordering and carrying cost as before i.e. Rs. 100 and Rs. 52, respectively. Adding Period 4 demand in the lot size:

Period:	1	2	3	4
Demand:	20	26	0	11
Lot Size:	57	0	0	0
Balance:	37	11	11	0

Again, ordering cost = Rs. 100
And carrying cost = 2 × (37 + 11 + 11) = 59 × 2 = Rs. 118

As the inventory holding cost is greater than the ordering cost, stop adding the demands further. Demands prior to this period (Period 4) i.e. up to Period 3 are clubbed to form the lot size, which is 46.

Starting from period 4 again, this process is repeated

Period:	4	4 + 5	4 + 5 + 6	4 + 5 + 6 + 7
Lot Size:	11	28	28	87
Ordering Cost (Rs.):	100	100	100	100
Carrying Cost (Rs.):	0	34	34	388

Period 7 is excluded because carrying cost exceeds ordering cost, and therefore:

Period:	4	5	6
Demand:	11	17	0
Lot Size:	28	0	0
Balance:	17	0	0

Now, from period 7 onwards, LTC method is applied and the complete schedule is obtained as follows.

Period:	1	2	3	4	5	6	7	8
Demand:	20	26	0	11	17	0	59	13
Lot Size:	46	0	0	28	0	0	72	0
Balance:	26	0	0	17	0	0	13	0

As three orders with lot size 46, 28 and 72 are placed, total ordering cost = 3 × 100 = Rs. 300. Inventory carrying cost is 2 × (26 + 17 + 13) = 2 × 56 = Rs. 112. Total cost of (300 + 112) = Rs. 412 corresponds to the schedule obtained using LTC method.

5.1.3 Least Unit Cost (LUC)

The total cost is obtained after considering the successive period demands and the total cost is divided by the lot size under consideration in order to obtain cost per unit item. Demands for the successive period are added until a decrease in total cost per unit is observed. Stopping criteria would be the increase in total cost per unit, and demands up to prior period than that are clubbed to form lot size.

In the previous example, first period demand is 20 and total cost corresponding to this lot size is 100 and therefore total cost per unit = 100/20 = 5.

Similarly, next period demand, i.e. 26, is added to 20 and lot size is 46 and total cost is (100 + 52) = Rs. 152. Total cost per unit = 152/46 = 3.30. The process is continued, as shown following:

Period:	**1**	**1 + 2**	**1 + 2 + 3**	**1 + 2 + 3 + 4**
(Under Consideration)				
Lot size:	20	46	46	57
Total Cost (Rs.):	100	152	152	218
Total Cost/Unit:	5	3.30	3.30	3.82

As with the lot size 57, per unit cost (Rs. 3.82) increases in comparison with the previous value (Rs. 3.30), Period 4 is excluded and three period demands are covered in the lot size.

The process is repeated form 4th period again, as follows:

Period:	**4**	**4 + 5**	**4 + 5 + 6**	**4 + 5 + 6 + 7**
(Under Consideration)				
Lot Size:	11	28	28	87
Total Cost (Rs.):	100	134	134	488
Cost/Unit:	9.09	4.78	4.78	5.60

Period 7 is excluded, as by including it, per unit cost is increased at Rs. 5.60. Now:

Period:	**7**	**7 + 8**
Lot Size:	59	72
Total Cost:	100	126
Cost/Unit:	1.69	1.75

Therefore, only Period 7 demand is considered as a lot.

The complete schedule obtained using LUC is:

Period:	**1**	**2**	**3**	**4**	**5**	**6**	**7**	**8**
Demand:	20	26	0	11	17	0	59	13
Lot Size:	46	0	0	28	0	0	59	13

Period:	1	2	3	4	5	6	7	8
Balance:	26	0	0	17	0	0	0	0

As four order are placed, ordering cost = 4 × 100 = Rs. 400.
Inventory carrying cost = 2 × (26 + 17) = Rs. 86.
Total cost = 400 + 86 = Rs. 486.

By using LUC, higher cost is obtained as compared to LTC.

5.1.4 Least Period Cost (LPC)

This is also called the Silver–Meal heuristic. Total cost is divided by number of periods covered by the lot size. The stopping criteria would be the increase observed in total cost per period. Consider the example of Section 5.1.3:

1.	**Period Under Consideration:**	**1**	**1 + 2**	**1 + 2 + 3**	**1 + 2 + 3 + 4**
2.	**Number of Periods Covered:**	1	2	3	4
3.	**Total Cost (Rs.):**	100	152	152	218
4.	**Total Cost Per Period = (c/b):**	100	76	50.67	54.5

Total cost was obtained using LUC before. It is divided by the number of periods covered. For example, grouping of first three period demands gives the total cost as Rs. 152, and 152/3 = 50.67 is the total cost per period. The least total cost per period is Rs. 50.67, because after that, it increases and the three period demands form the lot size.

From period 4 onwards:

1.	**Period Under Consideration:**	**4**	**4 + 5**	**4 + 5 + 6**	**4 + 5 + 6 + 7**
2.	**Number of Periods Covered:**	1	2	3	4
3.	**Total Cost (Rs.):**	100	134	134	488
4.	**Total Cost Per Period (c/b):**	100	67	44.67	122

Again three period demands are grouped.

Now:

1.	**Period:**	**7**	**8**
2.	**Number of Periods:**	1	2
3.	**Total Cost:**	100	126
4.	**Total Cost Per Period:**	100	63

As the total cost per period decreases by combining periods 7 and 8, obtained complete schedule is:

Period:	1	2	3	4	5	6	7	8
Demand:	20	26	0	11	17	0	59	13
Lot Size:	46	0	0	28	0	0	72	0
Balance:	26	0	0	17	0	0	13	0

Ordering cost = 3 × 100 = Rs. 300
Carrying cost = 2 × (26 + 17 + 13) = Rs. 112
Total cost = 300 + 112 = Rs. 412

5.1.5 Economic Ordering Quantity (EOQ)

Economic ordering quantity (EOQ) = $\sqrt{\dfrac{2rC}{F_h}}$

Where:

r = average demand per period
C = ordering cost
F_h = inventory carrying cost per unit per period

Let,

Period:	1	2	3	4	5	6	7	8
Demand:	20	26	0	11	17	0	59	13

Sum of the demands for all the eight periods is 146 and average demand per period:

$$r = \frac{146}{8} = 18.25$$

C = Rs. 200 and F_h = Rs. 2,

$$\text{then, EOQ} = \sqrt{\frac{2 \times 18.25 \times 200}{2}}$$
$$= 60.41 \approx 60$$

Using EOQ as lot size frequently,

Period:	1	2	3	4	5	6	7	8
Demand:	20	26	0	11	17	0	59	13
Lot Size:	60	0	0	0	60	0	60	0
Balance:	40	14	14	3.	46	46	47	34

A lot size of 60 units is ordered in the first period. At the end of Period 4, balance inventory is 3, which is less than the demand of Period 5, i.e. 17, and therefore EOQ is ordered in Period 5. Similarly in Period 7, 60 units are ordered. Ordering cost is (3 × 200) = Rs. 600, and carrying cost is Rs. 488 for the schedule.

Unlike other lot sizing methods, a balance inventory of 34 units remains at the end of last period, i.e. Period 8. However, the last order (during Period 7) may be adjusted so that no inventory exists at the end of planning period, as follows:

Period:	1	2	3	4	5	6	7	8
Demand:	20	26	0	11	17	0	59	13
Lot Size:	60	0	0	0	60	0	26	0
Balance:	40	14	14	3	46	46	13	0

Now the carrying cost is Rs. 352 as compared to that of Rs. 488 for the previous schedule.

5.1.6 Period Order Quantity (POQ)

Let,

Period:	1	2	3	4	5	6	7	8	9
Demand:	20	26	0	11	17	0	59	13	34

Total demand = 180, and average demand per period, $r = \frac{180}{9} = 20$.

To illustrate the POQ (period order quantity) method, assume ordering cost, C = Rs. 180 and inventory carrying cost per unit per period F_h = Rs. 2.

$$EOQ = \sqrt{\frac{2 \times 20 \times 180}{2}}$$

$$= 60$$

Number of periods to be covered by an order $= \frac{EOQ}{r} = \frac{60}{20}$

$$= 3 \text{ periods}$$

Therefore, three period demands are clubbed together in one order as shown following:

Period:	1	2	3	4	5	6	7	8	9
Demand:	20	26	0	11	17	0	59	13	34
Lot Size:	46	0	0	28	0	0	106	0	0
Balance:	26	0	0	17	0	0	47	34	0

The first three period demands constitute the lot size 46. The next three period (periods 4–6) demands are included in the lot size 28. Demands for periods 7–9 are 59, 13 and 34, respectively. The order to be placed in Period 7 is of (59 + 13 + 34) = 106 units.

Ordering cost = 3 × 180 = Rs. 540
Carrying cost = 2 × [26 + 17 + 47 + 34] = Rs. 248
Total cost = Rs. 788

The performance of different MRP lot sizing methods depend on factors such as:

1. Demand pattern
2. Parameters such as ordering/setup cost and inventory carrying cost
3. Planning periods

Few selected methods are applied for a given situation, and the schedule with good results is implemented in practice.

5.2 ALLOWING BACKORDERS

Consider the following example with 12 planning periods. EOQ is applied as MRP lot sizing method assuming that the initial and ending inventory are zero. As the total demand is 300, average demand $r = \frac{300}{12} = 25$,

$$\text{And EOQ} = \sqrt{\frac{2\times25\times150}{1.5}}$$
$$= 70.71 \approx 71 \text{ units}$$

using ordering cost, C = Rs. 150 and carrying cost per unit per period, F_h = Rs. 1.5.

Period:	**1**	**2**	**3**	**4**	**5**	**6**	**7**	**8**	**9**	**10**	**11**	**12**
Demand:	5	9	13	17	21	24	26	29	33	37	41	45
Lot Size:	71	0	0	0	0	71	0	71	0	71	0	16
Balance:	66	57	44	27	6	53	27	69	36	70	29	0

As five orders are placed, ordering cost = Rs. 750
Carrying cost = 484 × 1. 5 = Rs. 726
Total cost = Rs. 1,476.

Although shortages are generally not considered for deterministic demand situations in an MRP environment, sometimes these are unavoidable in the real world.

Let shortage cost per unit per period, K_h = Rs. 3
As all the shortages are assumed to be backordered,

$$\text{Optimal lot size, IOQ} = \sqrt{\frac{2rC(F_h+K_h)}{F_hK_h}}$$

This lot size is called integrated ordering quantity, IOQ.

$$\text{and IOQ} = \sqrt{\frac{2\times25\times150\times(1.5+3)}{1.5\times3}}$$
$$= 86.6 \approx 87 \text{ units}$$
$$\text{and optimal shortage quantity} = \sqrt{\frac{2rCF_h}{(F_h+K_h)K_h}}$$
$$= \sqrt{\frac{2\times25\times150\times1.5}{(1.5+3)\times3}}$$
$$= 28.86 \approx 29 \text{ units}$$

Using IOQ in the present example:

Period:	**1**	**2**	**3**	**4**	**5**	**6**	**7**	**8**	**9**	**10**	**11**	**12**
Demand:	5	9	13	17	21	24	26	29	33	37	41	45
Lot Size:	87	0	0	0	0	0	0	87	0	87	0	39
Balance:	82	73	60	43	22	−2	−28	30	−3	47	6	0

A total of 87 units are scheduled in the first period. If the balance falls below the optimal shortage quantity (29 units), then the IOQ is scheduled in that period. In

Period 7, balance is −28, and if 87 units are not scheduled in Period 8, then the balance falls below 29 units. Similarly, in Period 10, 87 units are scheduled. As the ending inventory is zero, the last lot size is of 39 units in Period 12.

Now ordering cost = 4 × 150 = Rs. 600

Carrying cost is based on the positive balance at the end of any period and is equal to 1.5 × (82 + 73 + 60 + 43 + 22 + 30 + 47 + 6) = Rs. 544.5

Shortage or backordering cost is based on negative balance in any period, equal to 3 × (2 + 28 + 3) = Rs. 99

Shortage quantity 2, 28 and 3 units are observed in periods 6, 7 and 9, respectively.

Total cost with this schedule = 600 + 544.5 + 99 = Rs. 1,243.50

The total cost is decreased by allowing backorders. The analysis is useful in the situations when any supplier has shown inability to send the desired quantity completely due to some temporary constraints. It helps in negotiation with the supplier as to how much shortage quantity and order quantity may be allowed in particular periods. It may be used in short-run rescheduling in spite of its limited applications.

5.3 PERIOD ORDER QUANTITY BASED ON INTEGRATED ORDERING QUANTITY (IOQ)

Period order quantity (POQ) method is discussed in Section 5.1.6. Similar logic is applied based on IOQ instead of EOQ to calculate the number of periods. A certain integer number of periods is determined to cover the demand, and shortages are not allowed. This heuristic is called integrated period order quantity (IPO).

Number of periods to be covered by an order = IOQ/r
In the previous example, r = 25 and IOQ = 87
and number of periods = 3.48 ≈ 3

Using three period demands as lot size:

Period:	1	2	3	4	5	6	7	8	9	10	11	12
Demand:	5	9	13	17	21	24	26	29	33	37	41	45
Lot Size:	27	0	0	62	0	0	88	0	0	123	0	0
Balance:	22	13	0	45	24	0	62	33	0	86	45	0

Total cost = ordering + carrying cost
= 600 + (1.5 × 330) = Rs. 1,095

In this heuristic, the backordering cost is used in order to compute the number of periods, but the backorders are not there in the schedule generated. In case, the estimation of shortage cost is not so precise; then also it will not affect the total system cost. If it gives lower costs than other mostly used lot sizing rules, then it may be applied.

Silver–Meal (SM) or least period cost is a promising and very popular technique. Using SM in the present example, following schedule is obtained:

Period:	1	2	3	4	5	6	7	8	9	10	11	12
Demand:	5	9	13	17	21	24	26	29	33	37	41	45
Lot Size:	27	0	0	38	0	50	0	62	0	78	0	45
Balance:	22	13	0	21	0	26	0	33	0	41	0	0

Total cost = Rs. 1,134

In the present example, total cost i.e. TC (IPO) is less than TC (SM).

With the increasing value of backordering cost Kh, IOQ tends towards EOQ and TC (IPO) would be equivalent to TC (POQ).

5.4 TEMPORARY PRICE DISCOUNTS/PRICE INCREASES

As discussed in Chapter 4, a price discount offer is applicable for short time and large quantities may be procured to take the advantage of this offer. Similarly, in the case of a declared price increase, a large quantity is purchased before the price increase becomes effective. EOQ is responsive enough for such situations, even in the MRP environment.

5.4.1 Temporary Price Reduction

Let the purchase cost per unit of the product be Rs. 10, and annual inventory holding cost fraction F is 0.25.

Annual inventory carrying cost per unit = 10 × 0.25 = Rs. 2.5

If there are 12 planning periods and each period represents a month, then inventory carrying cost in Rs. per unit per period,

$$F_h = \frac{2.5}{12}$$

$$EOQ = \sqrt{\frac{2rC}{F_h}}$$

Assume average demand, r = 25, and ordering cost, C = Rs. 150.

$$EOQ = \sqrt{\frac{2 \times 25 \times 150}{\left(\frac{2.5}{12}\right)}}$$

$$= 189.74 \approx 190$$

Applying the EOQ for constant demand pattern:

Period:	**1**	**2**	**3**	**4**	**5**	**6**	**7**	**8**	**9**	**10**	**11**	**12**
Demand:	25	25	25	25	25	25	25	25	25	25	25	25
Lot Size:	190	0	0	0	0	0	0	110	0	0	0	0
Balance:	165	140	115	90	65	40	15	100	75	50	25	0

As the total demand = 25 × 12 = 300,

Purchase cost = 300 × 10 = Rs. 3,000

Ordering cost = Rs. 300

Carrying cost = $880 \times \frac{2.5}{12}$ = Rs. 183.33

Total cost = Rs. 3,483.33

Now, short-term price discount offer is assumed to be applicable at the start of Period 1 and special lot size may be procured in Period 1 at purchase cost per unit of Rs. 9 if a 10% price discount is offered. The special lot size is obtained by Equation 4.22 in Section 4.1.2. As the initial and ending inventory is assumed to be zero, q = 0 and Equation 4.22 can be written as:

$$Q_a^* = \left(d + \frac{C}{Q^*}\right)\frac{R}{(P-d)F} + \frac{PQ^*}{2(P-d)}$$

Where:

d = Re. 1
R = 300
P = Rs. 10
Q* = 189.74
F = 0.25
C = Rs. 150

$$Q_a^* = \left(1 + \frac{150}{189.74}\right)\frac{300}{9 \times 0.25} + \frac{10 \times 189.74}{2 \times 9} = 343.78$$

Although special lot size is not obtained for MRP situation, it provides a base for use in this environment when price discount exists. As the Qa* is greater than the maximum requirement of 300 in this example, all 300 units may be procured in Period 1.

Period:	**1**	**2**	**3**	**4**	**5**	**6**	**7**	**8**	**9**	**10**	**11**	**12**
Demand:	25	25	25	25	25	25	25	25	25	25	25	25
Lot Size:	300	0	0	0	0	0	0	0	0	0	0	0
Balance:	275	250	225	200	175	150	125	100	75	50	25	0

Purchase cost = 300 × 9 = Rs. 2,700
Ordering cost = Rs. 150

Carrying cost = 1,650 × $\frac{9 \times 0.25}{12}$ = Rs. 309.38
Total cost = Rs. 3,159.38
Thus, a saving of (3,483.33 – 3,159.38) = Rs. 323.95

Using SM heuristic in the price discount situation:

Period:	**1**	**2**	**3**	**4**	**5**	**6**	**7**	**8**	**9**	**10**	**11**	**12**
Demand:	25	25	25	25	25	25	25	25	25	25	25	25
Lot Size:	200	0	0	0	0	0	0	0	100	0	0	0
Balance:	175	150	125	100	75	50	25	0	75	50	25	0

Ordering cost = Rs. 300

$$\text{Carrying cost} = \left(700 \times \frac{9 \times 0.25}{12}\right) + \left(150 \times \frac{10 \times 0.25}{12}\right) = \text{Rs. } 162.50$$

Purchase cost = (200 × 9) + (100 × 10) = Rs. 2,800
Total cost = Rs. 3,262.50

In the present example, cost benefit is greater when making use of the special lot size.

5.4.2 Price Increases

A price increase of k is declared and an additional quantity is procured just after the start of first period, i.e. before the price increase takes effect.

Consider the previous example in which,

P = Rs. 10, R = 300, F = 0.25 and C = Rs. 150
Let k = Re. 1

For using SM to determine first lot size, effective price per unit is Rs. 10 and carrying cost per unit per period is (10 × 0.25)/12. The schedule is:

Period:	**1**	**2**	**3**	**4**	**5**	**6**	**7**	**8**	**9**	**10**	**11**	**12**
Demand:	25	25	25	25	25	25	25	25	25	25	25	25
Lot Size:	200	0	0	0	0	0	0	0	100	0	0	0
Balance:	175	150	125	100	75	50	25	0	75	50	25	0

$$\text{Carrying cost} = \left(700 \times \frac{10 \times 0.25}{12}\right) + \left(150 \times \frac{11 \times 0.25}{12}\right) = \text{Rs. } 180.21$$

Purchase cost = (200 × 10) + (100 × 11) = Rs. 3,100
Ordering cost = 2 × 150 = Rs. 300
Total cost = Rs. 3,580.21

Using special lot size given by Equation 4.38 in Section 4.3.2 with q = 0:

$$Q_a^* = \frac{kR}{PF} + \frac{1}{P}\sqrt{\frac{2CR(P+k)}{F}}$$

$$= \frac{1 \times 300}{10 \times 0.25} + \frac{1}{10}\sqrt{\frac{2 \times 150 \times 300 \times 11}{0.25}}$$

$$= 318.99$$

As this is more than the total requirement of 300, the schedule is:

Period:	**1**	**2**	**3**	**4**	**5**	**6**	**7**	**8**	**9**	**10**	**11**	**12**
Demand:	25	25	25	25	25	25	25	25	25	25	25	25
Lot Size:	300	0	0	0	0	0	0	0	0	0	0	0
Balance:	275	250	225	200	175	150	125	100	75	50	25	0

$$\text{Carrying cost} = 1{,}650 \times \frac{10 \times 0.25}{12} = \text{Rs. } 343.75$$

Purchase cost = 300 × 10 = Rs. 3,000
Ordering cost = Rs. 150
Total cost = Rs. 3,493.75

The performance of different lot sizing techniques may depend on factors such as planning period, percentage price increase/decrease, different type of costs and demand patterns. However, the special lot size is responsive enough to be considered in the situation of temporary price discount and declared price increase.

Exercises

1. What do you understand by the master production schedule (MPS)?
2. Explain material requirements planning (MRP).
3. Assume the following net requirements:

Period:	**1**	**2**	**3**	**4**	**5**	**6**	**7**	**8**	**9**
Net Demand:	35	46	39	41	47	21	23	17	24

If the company procures for the demand of three periods, generate the procurement schedule in the following form:

Period:	**1**	**2**	**3**	**4**	**5**	**6**	**7**	**8**	**9**
Demand:									
Procurement Schedule:									

4. What do you understand by the MRP lot sizing?
5. Assume the information as follows:

Period:	**1**	**2**	**3**	**4**	**5**	**6**	**7**	**8**
Net Demand:	18	0	19	0	19	0	47	11

Implement LFL in the following form and calculate the total cost if setup/ordering cost is Rs. 40:

Period:	**1**	**2**	**3**	**4**	**5**	**6**	**7**	**8**
Net Demand:								
Production/ Procurement Schedule:								
Balance:								

6. Consider the following demand data:

Period:	1	2	3	4	5	6	7	8
Demand:	19	25	0	13	15	0	44	9

Obtain the complete schedule in the following form using LTC:

Period:	1	2	3	4	5	6	7	8
Demand:								
Lot Size:								
Balance:								

Let ordering cost be Rs. 120 and carrying cost be Rs. 3/unit period. Also find out the total cost with the suggested schedule.

7. Relevant information is given following:

Period:	1	2	3	4	5	6	7	8
Demand:	21	23	0	17	14	0	61	16

Implementing LUC, generate the plan in the form as follows:

Period:	1	2	3	4	5	6	7	8
Demand:								
Lot Size:								
Balance:								

Find out the related total cost of the generated plan if ordering cost is Rs. 135 and carrying cost is Rs. 2/unit period.

8. Following are the input details:

Period:	1	2	3	4	5	6	7	8
Demand:	34	14	0	28	25	0	38	40

What will be your approach for implementing 'least period cost' for these data? If ordering cost is Rs. 150 and carrying cost is Rs. 3/unit period, evaluate the total cost of the proposed plan, which is projected in the following form:

Period:	1	2	3	4	5	6	7	8
Demand:								
Lot Size:								
Balance:								

9. Various cost parameters are as follows:

   Ordering cost = Rs. 180
   Inventory holding cost = Rs. 2 per unit per period

   With the above cost parameters, utilize EOQ for the following situation and find out the total cost:

Period:	**1**	**2**	**3**	**4**	**5**	**6**	**7**	**8**	**9**
Demand:	19	27	0	12	16	0	56	17	33

10. Consider Exercise 9. Implement POQ and evaluate the relevant costs for the suggested procurement/production plan.
11. The following exercise relates to 12 planning periods:

Period:	**1**	**2**	**3**	**4**	**5**	**6**	**7**	**8**	**9**	**10**	**11**	**12**
Demand:	6	8	12	18	22	23	24	31	32	38	40	46

   Additionalinformation is as follows:

   Ordering cost = Rs. 150
   Inventory carrying cost per unit- period = Rs. 1.5

   a. Implement EOQ and evaluate the total cost.
   b. All the shortages are assumed to be backordered and shortage cost per unit per period is Rs. 3.

      i. Calculate the appropriate ordering quantity
      ii. Find out the optimal shortage quantity
      iii. Generate the plan allowing backorders and evaluate the total cost

12. Consider Exercise 11b. Implement POQ on the basis of the calculated appropriate ordering quantity and evaluate the total cost for the revised plan.
13. In the context of: (a) temporary price reduction, and (b) declared price increase, comment on the following statement: "Considering certain cases, EOQ is also responsive in such situations in the MRP environment".
14. Consider Exercise 13. Explain the computations, giving suitable examples for:

   a. Temporary price reduction
   b. Declared price increase

6 Multiproduct Manufacturing

The present chapter discusses the manufacturing of several items. It is convenient to implement the policy of producing each item in each cycle of production such that all the items are manufactured in every cycle. The problem relates to the determination of optimum common cycle time, subject to the constraint that the sum of production time of all the items should not exceed the cycle time.

6.1 DETERMINATION OF OPTIMUM CYCLE TIME

Let there be n items in a family production environment. This is shown in Figure 6.1.

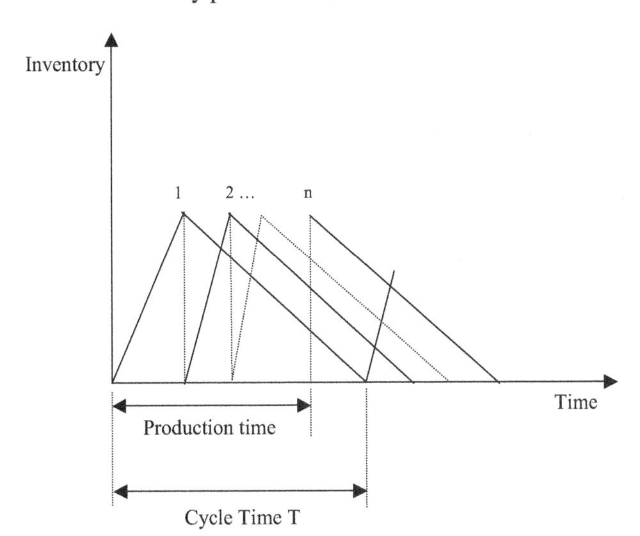

FIGURE 6.1 Family production environment.

Now:

A_i = Setup cost corresponding to an item i in Rs.
D_i = Annual demand of an item i
H_i = Inventory carrying cost for an item i in Rs./unit-year
P_i = Yearly production rate for an item i
Q_i = Production lot size for item i

DOI: 10.1201/9781003213994-6

For any item i as shown in Figure 6.2, maximum inventory:

$$V_i = \frac{Q_i(P_i - D_i)}{P_i} \tag{6.1}$$

$$\text{Cycle time } T = \frac{V_i}{(P_i - D_i)} + \frac{V_i}{D_i}$$

Using Equation 6.1, $T = \frac{Q_i}{D_i}$

$$\text{Or } Q_i = TD_i \tag{6.2}$$

$$\text{Production time} = \frac{Q_i}{P_i} = \frac{TD_i}{P_i} \tag{6.3}$$

As the sum of production time of all the n items should be less than the cycle time T,

$$\sum_{i=1}^{n} \frac{TD_i}{P_i} < T$$

$$\text{Or } \sum_{i=1}^{n} \frac{D_i}{P_i} < 1 \tag{6.4}$$

In order to have a feasible schedule, item parameters, viz. demand rate D_i and production rate P_i must satisfy the Condition 6.4.

Now total relevant cost consists of setup and inventory carrying cost.

Carrying cost for an item i,

$$= \frac{V_i}{2} \cdot H_i$$

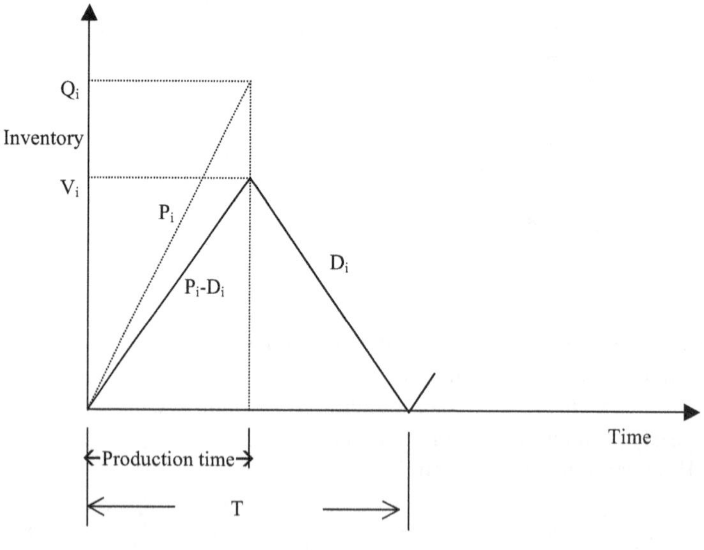

FIGURE 6.2 Production and consumption of an item i.

$$= \frac{Q_i(P_i - D_i)H_i}{2P_i}, \text{ using Equation 6.1}$$

$$= \frac{TD_iH_i(P_i - D_i)}{2P_i}, \text{ using Equation 6.2}$$

$$= \frac{T}{2} \cdot D_iH_i\left(1 - \frac{D_i}{P_i}\right)$$

And for n items, annual carrying cost $= \frac{T}{2} \cdot \sum_{i=1}^{n} D_iH_i\left(1 - \frac{D_i}{P_i}\right)$ (6.5)

For an item i, number of setups in one year $= \frac{D_i}{Q_i}$

$$= \frac{1}{T} \text{ using Equation 6.2}$$

And setup cost $= \frac{A_i}{T}$

$$\text{Annual setup cost for n items} = \frac{1}{T}\sum_{1}^{n} A_i \qquad (6.6)$$

By adding Equation 6.5 and Equation 6.6,

$$\text{Total relevant cost, } E = \frac{T}{2}\sum D_iH_i\left(1 - \frac{D_i}{P_i}\right) + \frac{1}{T}\sum A_i \qquad (6.7)$$

Differentiating with respect to T and equating to zero shows,

$$\text{Optimal } T = \sqrt{\frac{2\sum A_i}{\sum D_iH_i\left(1 - \frac{D_i}{P_i}\right)}} \qquad (6.8)$$

Example 6.1

Let there be two items, i.e. n = 2. Different parameters for these two items are as follows:

	Item i	
	1	**2**
Ai (Rs.)	100	150
Di (Units/Year)	400	300
Hi (Rs./Unit-Year)	7	5
Pi (Units/Year)	800	750

It may be verified that $\sum \frac{D_i}{P_i} < 1$

Now $D_1 H_1 \left(1 - \frac{D_1}{P_1}\right) = 400 \times 7 \left(1 - \frac{400}{800}\right) = 1400$

$$D_2 H_2 \left(1 - \frac{D_2}{P_2}\right) = 300 \times 5 \left(1 - \frac{300}{750}\right) = 900$$

And $\sum D_i H_i \left(1 - \frac{D_i}{P_i}\right) = 1400 + 900 = 2300$

$$\sum A_i = A_1 + A_2 = 100 + 150 = 250$$

Substituting the values in Equation 6.8, optimal cycle time,

$$T = \sqrt{\frac{2 \times 250}{2300}}$$
$$= 0.466 \text{ years}$$

From Equation 6.7, the total optimal cost, E = Rs. 1,072.38. Other results of interest such as optimum lot sizes may also be obtained using Equation 6.2:

$Q_1 = TD_1 = 0.466 \times 400 = 186.4$ units
$Q_2 = TD_2 = 0.466 \times 300 = 139.8$ units

6.2 ALLOWING BACKLOG

This is shown in Figure 6.3 for an item i. In the multiproduct environment, backorders may be allowed due to a variety of reasons such as manufacturing resource constraints, input item shortages, etc. The shortages are assumed to be completely backordered.

Let:

J_i = shortage quantity for item i
K_i = shortage or backordering cost per unit-year (for item i)

For an item i, the period (in years) during which shortages occur in a cycle

$$= \frac{J_i}{(P_i - D_i)} + \frac{J_i}{D_i} = \frac{J_i P_i}{D_i(P_i - D_i)}$$

As the average quantity is (J_i/2) and annual number of cycles are (1/T),

$$\text{Annual backordering/shortage cost} = \frac{J_i}{2} \cdot \frac{J_i P_i}{D_i(P_i - D_i)} \cdot K_i \cdot \frac{1}{T} = \frac{J_i^2 K_i}{2D_i \left(1 - \frac{D_i}{P_i}\right) T} \quad (6.9)$$

The period (in year) during which positive inventory exists,

$$= T - \frac{J_i}{D_i(1 - D_i / P_i)}$$

$$\text{And holding cost} = \frac{V_i}{2}\left[T - \frac{J_i}{D_i\left(1 - \frac{D_i}{P_i}\right)}\right]\frac{H_i}{T}$$

$$\text{As } \frac{V_i + J_i}{(P_i - D_i)} = \frac{Q_i}{P_i} = \frac{TD_i}{P_i},$$

$$V_i = \frac{(P_i - D_i)TD_i}{P_i} - J_i = TD_i\left(1 - \frac{D_i}{P_i}\right) - J_i$$

$$\text{And holding cost} = \frac{TD_iH_i\left(1 - \frac{D_i}{P_i}\right)}{2} + \frac{H_iJ_i^2}{2TD_i\left(1 - \frac{D_i}{P_i}\right)} - H_iJ_i \quad (6.10)$$

$$\text{Setup cost for an item i} = \frac{A_i}{T} \quad (6.11)$$

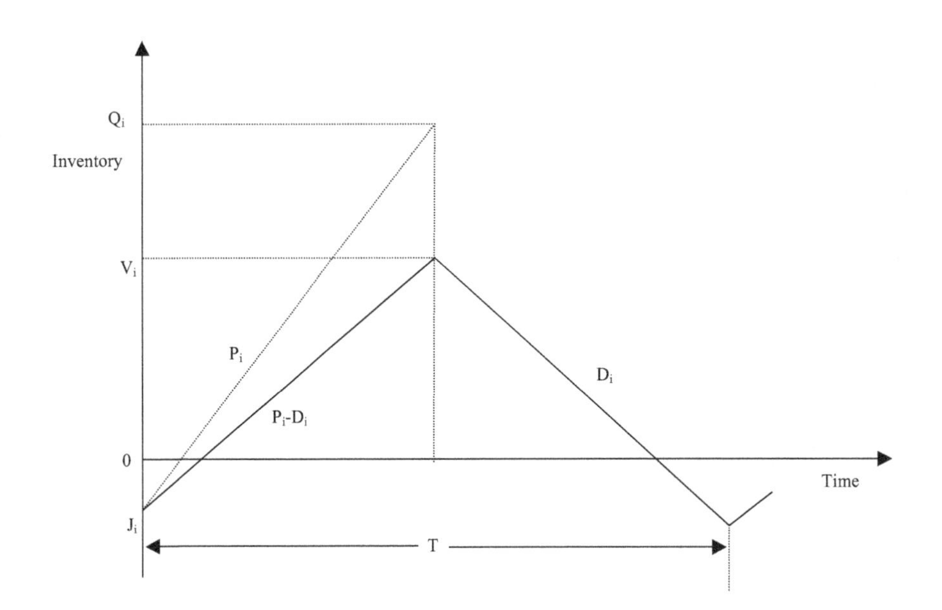

FIGURE 6.3 Allowing backorders in production system.

Total relevant cost consists of setup, holding and backordering cost. Adding Equations 6.9–6.11 for a family of n items, the total cost,

$$E = \frac{\Sigma A_i}{T} + \frac{T}{2}\Sigma D_i H_i\left(1-\frac{D_i}{P_i}\right) + \frac{1}{2T}\Sigma\frac{(H_i+K_i)J_i^2}{D_i\left(1-\frac{D_i}{P_i}\right)} - \Sigma H_i J_i \tag{6.12}$$

As this equation is convex in terms of each J_i, i=1,2, . . . n and T, differentiating with respect to each J_i and equating to zero, shows

$$\text{Optimal } J_i = \frac{TD_iH_i\left(1-\frac{D_i}{P_i}\right)}{(H_i+K_i)}, \text{ i=1,2, . . . n} \tag{6.13}$$

Substituting in Equation 6.12:

$$E = \frac{\Sigma A_i}{T} + \frac{T}{2}\Sigma\frac{D_iH_iK_i\left(1-\frac{D_i}{P_i}\right)}{(H_i+K_i)} \tag{6.14}$$

$\frac{\partial E}{\partial T} = 0$ shows

$$\text{Optimal } T = \sqrt{\frac{2\Sigma A_i}{\Sigma D_iH_iK_i\left(1-\frac{D_i}{P_i}\right)/(H_i+K_i)}} \tag{6.15}$$

Substituting in Equation 6.14 and Equation 6.13, respectively, optimum cost and shortages are obtained.

Example 6.2

Consider the data for Example 6.1. Additionally assume the annual shortage cost per unit as K_1 = Rs. 60 and K_2 = Rs. 70.

$$\text{Now } \frac{D_1H_1K_1\left(1-\frac{D_1}{P_1}\right)}{(H_1+K_1)} = 1253.73$$

$$\text{and } \frac{D_2H_2K_2\left(1-\frac{D_2}{P_2}\right)}{(H_2+K_2)} = 840$$

From (6.15), Optimal cycle time $T = \sqrt{\frac{2 \times 250}{(1253.73 + 840)}}$

$= 0.489$ year

From Equation 6.14, total optimal cost = Rs. 1,023.16.

Optimal shortage quantity obtained from (6.13),

$$J_1 = \frac{0.489 \times 400X7 \times \left(1 - \frac{400}{800}\right)}{(7 + 60)} = 10.22 \text{ units}$$

And $$J_2 = \frac{TD_2H_2\left(1 - \frac{D_2}{P_2}\right)}{(H_2 + K_2)} = 5.87 \text{ units}$$

Optimal production lot size from Equation 6.2,

$Q_1 = TD_1 = 195.6$ units
$Q_2 = TD_2 = 146.7$ units

Comparing with the results of Example 6.1, production quantity Q_i are increased to deal with the shortages.

By allowing shortages, decrease in cost = 1,072.38 – 1,023.16 = Rs. 49.22.

Substituting Equation 6.8 in Equation 6.7,

$$\text{Cost without backordering} = \sqrt{2(\Sigma A_i)\left[\Sigma D_iH_i\left(1 - \frac{D_i}{P_i}\right)\right]} \qquad (6.16)$$

Substituting Equation 6.15 in Equation 6.14, cost using optimal results with allowable backordering

$$= \sqrt{2(\Sigma A_i)\left[\Sigma D_iH_iK_i\left(1 - \frac{D_i}{P_i}\right) / (H_i + K_i)\right]} \qquad (6.17)$$

Subtracting Equation 6.17 from Equation 6.16, decrease in cost by allowing backorders:

$$= \sqrt{2(\Sigma A_i)}\left[\begin{array}{l}\sqrt{\Sigma D_iH_i(1 - D_i / P_i)} - \\ \sqrt{\left\{\Sigma D_iH_iK_i\left(1 - \frac{D_i}{P_i}\right) / (H_i + K_i)\right\}}\end{array}\right] \qquad (6.18)$$

As the value of $K_i/(H_i+K_i)$ is always less than 1, Expression 6.18 will always be positive and therefore decrease in the cost is observed by allowing backorders. However, care must be taken in estimating the shortage cost precisely. Further, the perception of an organization regarding satisfactory service level should also be considered in the decision making.

6.3 DIFFERENT CYCLE TIME FOR DIFFERENT ITEMS

Instead of common cycle time, if different C.T., i.e., T_i, i = 1,2, . . . n are used, then the total cost Equation 6.12 may be written as,

$$E = \Sigma\left(\frac{A_i}{T_i}\right) + \frac{1}{2}\Sigma T_i D_i H_i \left(1 - \frac{D_i}{P_i}\right) + \frac{1}{2}\Sigma \frac{(H_i + K_i)J_i^2}{T_i D_i \left(1 - \frac{D_i}{P_i}\right)} - \Sigma H_i J_i \tag{6.19}$$

Considering each J_i, $\frac{\partial E}{\partial J_i} = 0$ shows

$$\text{Optimal } J_i = \frac{T_i D_i H_i \left(1 - \frac{D_i}{P_i}\right)}{(H_i + K_i)} \tag{6.20}$$

Substituting in Equation 6.19,

$$E = \Sigma \frac{A_i}{T_i} + \frac{1}{2}\Sigma \frac{T_i D_i H_i K_i \left(1 - \frac{D_i}{P_i}\right)}{(H_i + K_i)} \tag{6.21}$$

Considering each T_i, $\frac{\partial E}{\partial T_i} = 0$ shows

$$\text{Optimal } T_i = \sqrt{\frac{2A_i(H_i + K_i)}{D_i H_i K_i \left(1 - \frac{D_i}{P_i}\right)}}, \ i=1,2, \ldots n \tag{6.22}$$

Substituting in Equation 6.21,

$$\text{Optimal } E = \Sigma \sqrt{\frac{2A_i D_i H_i K_i \left(1 - \frac{D_i}{P_i}\right)}{(H_i + K_i)}} \tag{6.23}$$

Example 6.3

Results are obtained using input parameters of Example 6.2:

From Equation 6.22,
$$T_1 = \sqrt{\frac{2A_1(H_1+K_1)}{D_1H_1K_1\left(1-\frac{D_1}{P_1}\right)}} = \sqrt{\frac{2\times 100}{1253.73}} = 0.399$$
year

And
$$T_2 = \sqrt{\frac{2A_2(H_2+K_2)}{D_2H_2K_2\left(1-\frac{D_2}{P_2}\right)}} = \sqrt{\frac{2\times 150}{840}} = 0.598 \text{ year}$$

From Equation 6.20,

$$J_1 = \frac{T_1D_1H_1\left(1-\frac{D_1}{P_1}\right)}{(H_1+K_1)} = \frac{0.399\times 400\times 7\left(1-\frac{400}{800}\right)}{(7+60)} = 8.34 \text{ units}$$

And
$$J_2 = \frac{T_2D_2H_2\left(1-\frac{D_2}{P_2}\right)}{(H_2+K_2)} = \frac{0.598\times 300\times 5\left(1-\frac{300}{750}\right)}{(5+70)} = 7.18 \text{ units}$$

Now
$$\frac{2A_1D_1H_1K_1\left(1-\frac{D_1}{P_1}\right)}{(H_1+K_1)} = 2\times 100\times 1253.73 = 250746$$

And
$$\frac{2A_2D_2H_2K_2\left(1-\frac{D_2}{P_2}\right)}{(H_2+K_2)} = 2\times 150\times 840 = 252000$$

From Equation 6.23, the optimal total cost $E = \sqrt{250746} + \sqrt{252000}$

= Rs. 1,002.74

Lower cost is obtained as compared to the common cycle time approach. This is because individual item parameters are considered separately in order to obtain an optimized individual cycle time. However, this strategy using different cycle time may not be convenient to implement, particularly when the number of items is greater in the family production context.

6.3.1 Without Stock Outs

When shortages are not allowed, Equation 6.7 is written using different cycle time,

$$E = \frac{1}{2}\Sigma T_i D_i H_i\left(1 - \frac{D_i}{P_i}\right) + \Sigma\frac{A_i}{T_i} \tag{6.24}$$

Differentiating with respect to each T_i and equating to zero yields,

$$\text{Optimal } T_i = \sqrt{\frac{2A_i}{D_i H_i\left(1 - \frac{D_i}{P_i}\right)}} \tag{6.25}$$

$$\text{Or } D_i H_i\left(1 - \frac{D_i}{P_i}\right) = \frac{2A_i}{T_i^2}$$

Substituting in Equation 6.24,

$$\text{Optimal } E = 2\Sigma\frac{A_i}{T_i} \tag{6.26}$$

In Example 6.1, common cycle time T obtained is 0.466 year and total cost is Rs. 1,072.38. Using the similar input data with different cycle time approach, the optimal results are:

From Equation 6.25, $T_1 = 0.378$ year and $T_2 = 0.577$ year
From Equation 6.26, E = 2[(100/0.378) + (150/0.577)] = Rs. 1,049.03

As expected, lower cost is obtained than with the common cycle time approach.

6.4 FRACTIONAL BACKLOGGING

In such situation, a fraction b_i of the shortage quantity is not backordered. Revisit Section 6.2 concerning the common cycle time approach.

$$\text{Annual shortage quantity for an item i} = \frac{J_i^2}{2D_i\left(1 - \frac{D_i}{P_i}\right)T}$$

$$\text{And annual shortage cost} = \frac{K_i J_i^2}{2D_i\left(1 - \frac{D_i}{P_i}\right)T} \tag{6.27}$$

As discussed in Chapter 2, K_i is the estimated shortage cost which is associated with all the shortages, whether these are backordered or not.

$$\text{Quantity which is not backlogged yearly} = \frac{b_i J_i^2}{2D_i\left(1 - \frac{D_i}{P_i}\right)T}$$

Where b_i = a fraction of shortage quantity which is not backordered for an item i.

In the case of fractional backordering, production time cost will also need to be included.

Let C = production time cost per year (in Rs./year)

Refer to Figure 6.3, ignoring b_i.

$$\text{Production time in a cycle} = \frac{Q_i}{P_i} = \frac{TD_i}{P_i}$$

As $\left(\frac{1}{T}\right)$ number of cycles are there per year, production time per year is $\left(\frac{D_i}{P_i}\right)$ and annual production cost for item i = $\frac{D_i}{P_i} \cdot C$

Annual production cost for item i with partial backordering:

$$= \frac{C}{P_i}\left(D_i - \frac{b_i J_i^2}{2D_i\left(1-\frac{D_i}{P_i}\right)T}\right)$$

$$= \frac{CD_i}{P_i} - \frac{Cb_i J_i^2}{2P_i D_i\left(1-\frac{D_i}{P_i}\right)T} \tag{6.28}$$

Holding cost for an item I yearly, is given by Equation 6.10 in Section 6.2 as:

$$= \frac{TD_i H_i\left(1-\frac{D_i}{P_i}\right)}{2} + \frac{H_i J_i^2}{2TD_i\left(1-\frac{D_i}{P_i}\right)} - H_i J_i \tag{6.29}$$

$$\text{Setup cost} = \frac{A_i}{T} \tag{6.30}$$

Adding Equations 6.27–6.30 for a group of n items, the total cost,

$$E = \frac{1}{2T}\Sigma\frac{(H_i+K_i)J_i^2}{D_i\left(1-\frac{D_i}{P_i}\right)} + C\Sigma\frac{D_i}{P_i} - \frac{C}{2T}$$
$$\Sigma\frac{b_i J_i^2}{P_i D_i\left(1-\frac{D_i}{P_i}\right)} + \frac{T}{2}\Sigma D_i H_i\left(1-\frac{D_i}{P_i}\right) - \Sigma H_i J_i + \frac{1}{T}\Sigma A_i \tag{6.31}$$

Differentiating partially with respect to each J_i and equating to zero, shows:

$$J_i = \frac{TD_i H_i\left(1-\frac{D_i}{P_i}\right)}{(H_i + K_i - Cb_i / P_i)}, \text{ i} = 1,2, \ldots \text{n} \tag{6.32}$$

Substituting in Equation 6.31:

$$E = C\sum\frac{D_i}{P_i} + \frac{1}{T}\sum A_i + \frac{T}{2}\sum\frac{D_i H_i\left(K_i - \frac{Cb_i}{P_i}\right)\left(1-\frac{D_i}{P_i}\right)}{\left(H_i + K_i - \frac{Cb_i}{P_i}\right)} \tag{6.33}$$

$\frac{\partial E}{\partial T} = 0$ shows

$$T = \sqrt{\frac{2\sum A_i}{\sum D_i H_i\left(K_i - \frac{Cb_i}{P_i}\right)\left(1-\frac{D_i}{P_i}\right) / \left(H_i + K_i - \frac{Cb_i}{P_i}\right)}} \tag{6.34}$$

A feasibility/optimality condition would be:

$$K_i > \frac{Cb_i}{P_i}$$

As $\left(\frac{C}{P_i}\right)$ is the unit production cost for item i, b_i is a fraction and unit shortage cost is greater than the unit production cost in the real world, the condition is easily satisfied in the practical environment. However, this may be verified from the input data set.

It can be shown that

$$\text{Optimal } E = C\sum\frac{D_i}{P_i} + \frac{2}{T}\sum A_i \tag{6.35}$$

Where T is given by Equation 6.34.

Example 6.4

Consider the following parameters:

Production time cost, C = Rs. 9,000 per year

	Item i	
	1	**2**
Ai (Rs.)	100	150
Di (Units/Year)	400	300
Hi (Rs./Unit-Year)	4	4
Pi (Units/Year)	800	750
Ki (Rs./Unit-Year)	120	140
bi	0.2	0.3

Now:

$$\frac{D_1H_1\left(K_1-\frac{Cb_1}{P_1}\right)\left(1-\frac{D_1}{P_1}\right)}{H_1+K_1-\frac{Cb_1}{P_1}}=$$

$$\frac{400\times 4\times\left[120-\left(\frac{9000\times 0.2}{800}\right)\right]\times 0.5}{4+120-\left(\frac{9000\times 0.2}{800}\right)}=773.72$$

Similarly:

$$\frac{D_2H_2\left(K_2-\frac{Cb_2}{P_2}\right)\left(1-\frac{D_2}{P_2}\right)}{H_2+K_2-\frac{Cb_2}{P_2}}=\frac{300\times 4\times(140-3.6)\times 0.6}{144-3.6}=699.49$$

From Equation 6.34, $T=\sqrt{\frac{2\times 250}{(773.72+699.49)}}=0.582$ year

From Equation 6.35, $E=9000\left(\frac{400}{800}+\frac{300}{750}\right)+\frac{2}{0.582}(100+150)$ = Rs. 8,959.11

From Equation 6.32, $J_1=\frac{0.582\times 400\times 4\times 0.5}{(124-2.25)}=3.82$ units

and $J_2=\frac{0.582\times 300\times 4\times 0.6}{(144-3.6)}=2.98$ units

6.5 SHELF LIFE

Shelf life refers to the time period during which an item may be stored before it gets spoiled. This may also include the factor of obsolescence. In case of seasonal items, their marketing gets affected if the items are stored beyond certain time period.

Referring to Figure 6.2, the production time is $\frac{TD_i}{P_i}$ given by Equation 6.3. Assuming the withdrawal of inventory item on the first-in-first-out (FIFO) basis, item i is stored for the time, $T - \frac{TD_i}{P_i}$ or $T\left(1 - \frac{D_i}{P_i}\right)$.

Let S_i = shelf life of an item i

$$\text{Shelf life constraint is } T\left(1 - \frac{D_i}{P_i}\right) \leq S_i \tag{6.36}$$

In Example 6.1, cycle time T is 0.466 year.

$$\frac{D_1}{P_1} = \frac{400}{800} = 0.5 \text{ and } \frac{D_2}{P_2} = \frac{300}{750} = 0.4$$

Assume that the shelf life, $S_1 = 0.21$ and $S_2 = 0.3$ year

$$\text{For item 1, } T\left(1 - \frac{D_1}{P_1}\right) \leq S_1 \text{ or } 0.466(1 - 0.5) \leq 0.21$$

As the L.H.S. is greater than S_1, shelf life constraint is not satisfied for Item 1. However for Item 2, $0.466(1 - 0.4) \leq 0.3$ or $0.2796 \leq 0.3$, and the shelf life constraint is satisfied for Item 2.

In order to implement common cycle time, Condition 6.36 is to be verified for each item. If any of the items in a family violates the shelf life constraint, T = 0.466 year cannot be used. It is to be reduced as per the requirement of the constrained Item 1. As Constraint 6.36 is a binding constraint,

$$\text{reduced cycle time } T_r = \frac{S_1}{\left(1 - \frac{D_1}{P_1}\right)} = 0.42 \text{ year}$$

Reduced cycle time, T_r, i.e. 0.42 year, is used as common cycle time in order to deal with the shelf life constraint. In case of more than one constrained item, the reduced cycle time T_r is evaluated for each such item and the smallest value of T_r is implemented.

The performance of different cycle times for different item is better in terms of cost, as discussed before. In the present situation also, this strategy performs better and should be followed if it is convenient to implement.

Example 6.5

Let the data for two items be similar to Example 6.1, with additional values of shelf lives.

	Item i	
	1	**2**
Ai (Rs.)	100	150
Di (Units/Year)	400	300
Hi (Rs./Unit-Year) 7	5	
Pi (Units/Year)	800	750
Si (Year)	0.21	0.3

To deal with the shelf life, apply both the approaches, viz. (a) common cycle time, and (b) different cycle time approach.

1. Common cycle time approach: In Example 6.1, evaluated optimal C.T. without shelf life consideration, T = 0.466 year and corresponding cost,

   E = Rs. 1,072.38.

Since Item 1 is having the constraint on shelf life,

$$\text{reduced cycle time } T_r = \frac{0.21}{\left(1-\frac{400}{800}\right)} = 0.42 \text{ year}$$

Substituting T_r, in place of T in Equation 6.7,

$$\text{Total cost } = \frac{T_r}{2}\Sigma D_i H_i\left(1-\frac{D_i}{P_i}\right)+\frac{1}{T_r}\Sigma A_i$$

$$= \frac{0.42}{2}\times 2300+\frac{1}{0.42}\times 250$$

$$= \text{Rs. } 1{,}078.24$$

2. Different cycle time approach: In Section 6.3.1, optimal results without shelf life consideration are:

$$T_1 = 0.378 \text{ year}, T_2 = 0.577 \text{ year and total cost} = \text{Rs. } 1{,}049.03.$$

From Equation 6.36, $T_1\left(1-\frac{D_1}{P_1}\right) \le S_1$ for item 1 which is satisfied.

For item 2, $T_2\left(1-\frac{D_2}{P_2}\right) \le S_2$, which is not satisfied.

Therefore, Item 2 is the constrained item in this approach.

$$\text{Reduced cycle time } T_2^1 = \frac{S_2}{\left(1-\frac{D_2}{P_2}\right)}$$

$$= \frac{0.3}{\left(1 - \frac{300}{750}\right)} = 0.5 \text{ year}$$

From Equation 6.24, total cost $= \frac{1}{2} \Sigma T_i D_i H_i \left(1 - \frac{D_i}{P_i}\right) + \Sigma \frac{A_i}{T_i}$

Now $T_1 D_1 H_1 \left(1 - \frac{D_1}{P_1}\right) = 529.2$

and $T_2^{\,1} D_2 H_2 \left(1 - \frac{D_2}{P_2}\right) = 450$

and total cost with shelf life $= \left(\frac{1}{2}\right)\left[529.2 + 450\right] + \left[\left(\frac{100}{0.378}\right) + \left(\frac{150}{0.5}\right)\right]$

= Rs. 1,054.15

Which is less than the corresponding cost with common C.T. approach (Rs. 1,078.24), as expected.

6.6 INCORPORATING INPUT ITEM PROCUREMENT

Multiproduct manufacturing environment is shown in Figure 6.4 along with input item procurement.

Let:

D_{iF} = annual demand for finished product i
H_{iF} = carrying cost for finished product i
P_{iF} = annual production rate for finished product i
A_{iF} = setup cost for finished product i

Total cost concerning production of a group of n end items is given by Equation 6.7. Writing it as:

$$= \frac{T}{2} \Sigma D_{iF} H_{iF} \left(1 - \frac{D_{iF}}{P_{iF}}\right) + \frac{1}{T} \Sigma A_{iF} \tag{6.37}$$

Smaller lot sizes of raw material or input item are planned in each manufacturing cycle. If the number of orders of the input item for finished product i per production cycle = k_i, then

Quantity of input item to be ordered $= \frac{TD_{iF}}{k_i}$

as the production quantity per cycle $= TD_{iF}$

(It is assumed that one unit of input item is required per unit of the end product. Each 'unit' of the input item may be considered as equivalent to the total requirements of raw materials per unit of the end product.)

If H_{iR} is the holding cost of input item corresponding to an end product i, then

Annual carrying cost for an input item $= \frac{TD_{iF}}{2k_i} \cdot \frac{D_{iF}}{P_{iF}} \cdot H_{iR} = \frac{TD_{iF}^2 H_{iR}}{2P_{iF}k_i}$

Because average inventory is $\frac{TD_{iF}}{2k_i}$ which exists for the time $\frac{D_{iF}}{P_{iF}}$ in a year.

Total cost concerning input items for a family of end items

$$= T\sum \frac{D_{iF}^2 H_{iR}}{2P_{iF}k_i} + \frac{1}{T}\sum k_i A_{iR} \tag{6.38}$$

Where A_{iR} = ordering cost of input item corresponding to finished item i

Adding Equation 6.37 and Equation 6.38, total relevant cost:

$$E = \frac{T}{2}\sum D_{iF}H_{iF}\left(1 - \frac{D_{iF}}{P_{iF}}\right) + \frac{1}{T}\sum A_{iF} + \frac{1}{T}\sum k_i A_{iR} + T\sum \frac{D_{iF}^2 H_{iR}}{2P_{iF}k_i} \tag{6.39}$$

As it is a convex function in terms of each k_i and T, differentiating partially with respect to each k_i and equating to zero gives

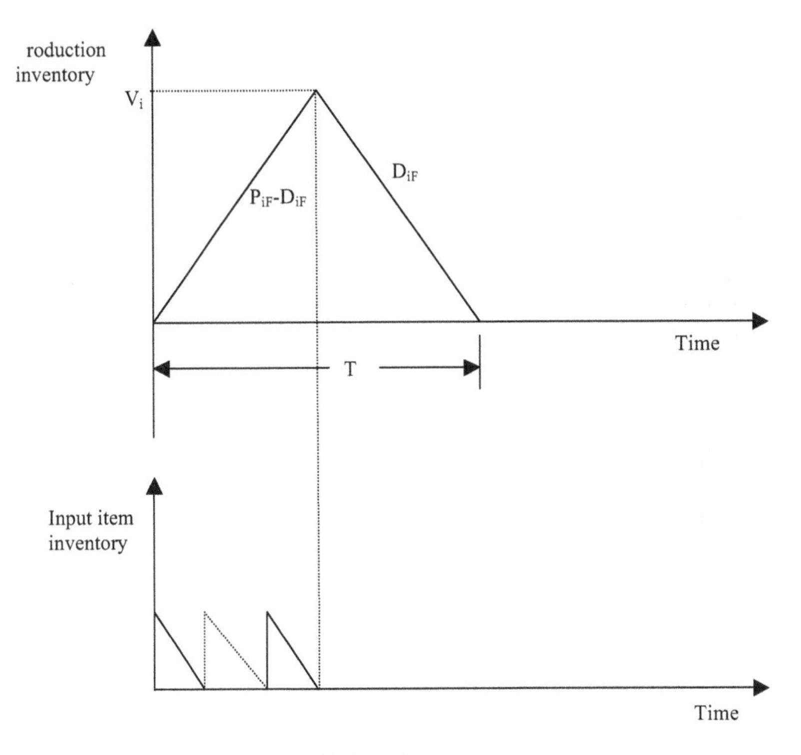

FIGURE 6.4 Manufacturing cycle with input item procurement.

$$k_i = TD_{iF}\sqrt{\frac{H_{iR}}{2P_{iF}A_{iR}}}, \text{ i} = 1,2, \ldots \text{n} \quad (6.40)$$

Substituting in Equation 6.39,

$$E = \frac{T}{2}\sum D_{iF}H_{iF}\left(1-\frac{D_{iF}}{P_{iF}}\right)+\frac{1}{T}\sum A_{iF}+\sum D_{iF}\sqrt{\frac{2H_{iR}A_{iR}}{P_{iF}}} \quad (6.41)$$

$\frac{\partial E}{\partial T} = 0$ shows optimal values as:

$$T^* = \sqrt{\frac{2\sum A_{iF}}{\sum D_{iF}H_{iF}\left(1-\frac{D_{iF}}{P_{iF}}\right)}} \quad (6.42)$$

And also it can be shown,

$$E^* = \frac{2\sum A_{iF}}{T^*}+\sum D_{iF}\sqrt{\frac{2H_{iR}A_{iR}}{P_{iF}}} \quad (6.43)$$

Example 6.6

Assume the following data:

	Item i	
	1	**2**
AiR (Rs.)	20	40
AiF (Rs.)	100	150
DiF (Units/Year)	400	300
HiR (Rs./Unit-Year)	6	4
HiF (Rs./Unit-Year)	7	5
PiF (Units/Year)	800	750

From Equation 6.42, T* = 0.466 year

From Equation 6.40, optimal values of k_i are:

$$k_1 = T^*D_{1F}\sqrt{\frac{H_{1R}}{2P_{1F}A_{1R}}} = 0.466\times 400\sqrt{\frac{6}{2\times 800\times 20}} = 2.55$$

And $k_2 = T^*D_{2F}\sqrt{\frac{H_{2R}}{2P_{2F}A_{2R}}} = 0.466\times 300\sqrt{\frac{4}{2\times 750\times 40}} = 1.14$

$$\text{Now } D_{1F}\sqrt{\frac{2H_{1R}A_{1R}}{P_{1F}}} = 400\sqrt{\frac{2\times 6\times 20}{800}} = 219.09$$

$$\text{And } D_{2F}\sqrt{\frac{2H_{2R}A_{2R}}{P_{2F}}} = 300\sqrt{\frac{2\times 4\times 40}{750}} = 195.96$$

From Equation 6.43, optimum total cost,

$$E^* = \frac{2(100+150)}{0.466} + (219.09+195.96)$$

$$= \text{Rs. } 1{,}488.01$$

6.7 FLEXIBILITY IN THE PRODUCTION RATE

In some cases, there is a need to change the production rate. Production rate may be decreased in order to have less inventory. To deal with the shelf life constraint, decrease in the production rate alone is an inferior option in most situations. However, this is one among many available options. There is a possibility of increasing or decreasing the production rate when simultaneous variation of cycle time and production rate is being explored. Otherwise also, to incorporate flexibility, altering the production rate is a matter of interest in variety of situations.

In Section 6.4, production time cost per year, C (in Rs./year) is used. But such fixed production time cost may not be useful when production rate is altered from the current level. The cost may vary as per the industry type or particular shop floor.

Generalized production cost of running a machine in Rs./year

$$= C\left(\frac{P_i}{D_i}\right)^{\alpha} \tag{6.44}$$

Where:

α = shop floor index
P_i = annual production rate
D_i = annual demand
C = fixed production time cost in Rs./year

When $\alpha = 0$, generalized production cost reduces to the fixed production cost C, which was used before.

Refer to Figure 6.2. Production time per cycle is given by Equation 6.3 as $\frac{TD_i}{P_i}$

$$\text{Production time per year} = \frac{1}{T}\cdot\frac{TD_i}{P_i} = \frac{D_i}{P_i}$$

Using generalized production cost in Rs./year,

$$\text{Annual production cost for item i} = C\left(\frac{P_i}{D_i}\right)^{\alpha}\left(\frac{D_i}{P_i}\right)$$

$$= C\left(\frac{D_i}{P_i}\right)^{1-\alpha} \tag{6.45}$$

Adding Equation 6.7 and Equation 6.45 for multiple items, the total relevant cost,

$$E = \frac{T}{2}\Sigma D_i H_i\left(1-\frac{D_i}{P_i}\right)+\frac{1}{T}\Sigma A_i + C\Sigma\left(\frac{D_i}{P_i}\right)^{1-\alpha} \tag{6.46}$$

It can be shown that the optimal values are:

$$T^* = \sqrt{\frac{2\Sigma A_i}{\Sigma D_i H_i\left(1-\frac{D_i}{P_i}\right)}} \tag{6.47}$$

$$\text{And } E^* = \frac{2}{T^*}\Sigma A_i + C\Sigma\left(\frac{D_i}{P_i}\right)^{1-\alpha} \tag{6.48}$$

Example 6.7

Let, C = Rs. 9,000 per year

$\alpha = 0.2$

	Item i	
	1	**2**
Ai (Rs.)	100	150
Di (Units/Year)	400	300
Pi (Units/Year)	800	750
Hi (Rs./Unit-Year)	7	5

From Equation 6.47, T* = 0.466 year

and from Equation 6.48, $E^* = \frac{2}{0.466}\times 250 + 9000\left[\left(\frac{400}{800}\right)^{0.8}+\left(\frac{300}{750}\right)^{0.8}\right]$

= Rs. 10,566.15

Similarly for different values of shop floor index, the total cost is evaluated as follows:

α :	0	0.2	0.4	0.6	0.8	1.0
E*(Rs.):	9,172.96	10,566.15	12,204.47	14,131.99	16,400.89	19,072.96

Total costs are more sensitive towards higher values of shop floor index α .

With reference to α = 0.2, cost obtained is Rs. 10,566.15, and production rate P_1 = 800 and P_2 = 750. Now at the similar level of α , production rate of both the items are increased by a certain percentage and the effects are shown below:

% Increase in Production Rate	P1	P2	T* (Year)	E*(Rs.)
10	880	825	0.449	9,910.22
20	960	900	0.436	9,352.25
30	1,040	975	0.425	8,870.93

Considering T* = 0.466 year and E* = Rs. 10,566.15 as reference, percentage decrease in T* and E* respectively are obtained as:

% Increase in Pi	% Decrease in T*	% Decrease in E*
10	3.65	6.21
20	6.44	11.49
30	8.80	16.04

Generalized production cost in Rs./year has increased by increasing the production rate, as it is given by $C\left(\frac{P_i}{D_i}\right)^{\alpha}$. But at the same time, production time per year, i.e. D_i/P_i, has reduced. Cycle time is decreased as the production rate is higher. Similarly, production rate of both the items are decreased and the computational results are shown as follows:

% Decrease in Production Rate	P1	P2	T* (Year)	E*(Rs.)
10	720	675	0.491	11,347.31
20	640	600	0.527	12,297.23
30	560	525	0.589	13,477.35

Again, reference values of T* and E* are 0.466 year and Rs. 10,566.15, respectively. Cycle time is increased as production rate is decreased. Sensitivity of cycle time and cost are as follows:

% Decrease in Pi	% Increase in T*	% Increase in E*
10	5.36	7.39
20	13.09	16.38
30	26.39	27.55

These data may give an indication of variation of optimal values. However, in decreasing the production rate, the care should be taken regarding feasibility of the data set. $\Sigma\left(\frac{D_i}{P_i}\right) < 1$, given by Equation 6.4, should be satisfied in order to have a feasible schedule. In the present example, decrease in the production rate of both the items may be up to 10% only; that, too, a limiting case. This is because,

$$\Sigma\left(\frac{D_i}{P_i}\right) = \left(\frac{400}{720}\right) + \left(\frac{300}{675}\right)$$
$$= 1$$

Considering a 10% increase as well as decrease in the production rate, the results are more sensitive for decrease in the production rate. Generalized production cost per year is decreased with the decrease in production rate, but production time has increased considerably, thereby increasing the total cost.

In this example, production rate of all the items are changed simultaneously. In practice, the change in the rate of production of one or more items in a family production context may be examined depending on the situation such as shelf life constraint. Decrease in the production rate for any item may also be due to certain resource constraint. Effects on the total cost are useful to make a decision in the matter, and also for short-term replanning.

Exercises

1. Explain the multi-item production environment with the help of Fig. Derive the following for such an environment when there is common cycle time:

   a. Annual carrying cost
   b. Annual setup cost
   c. Common cycle time

2. Various parameters for the two items are as follows:

	Item 1	Item 2
Setup cost (Rs.)	120	140
Demand (Units/Year)	320	380
Carrying Cost (Rs./Unit-Year)	6	8
Production Rate (Units/Year)	700	800

   Evaluate the optimum values of:

   a. Common cycle time
   b. Total cost
   c. Lot size for each item

3. Allow complete backordering in the multi-item production environment and derive:

a. Common cycle time
b. Total cost

4. Various parameters for the two items are as follows:

	Item 1	Item 2
Setup Cost (Rs.)	120	140
Demand (Units/Year)	320	380
Carrying Cost (Rs./Unit-Year)	8	7
Production Rate (Units/Year)	650	850
Annual Shortage Cost (Rs./Unit)	50	80

Evaluate the optimum values of:

a. Common cycle time
b. Total cost
c. Lot size for each item
d. Shortage quantity for each item

5. Allow fractional backordering in the multi-item production environment and derive:

a. Common cycle time
b. Total cost

6. Various parameters for the two items are as follows:
Production time cost per year = Rs. 9,500

	Item 1	Item 2
Setup Cost (Rs.)	135	125
Demand (Units/Year)	320	390
Carrying Cost (Rs./Unit-Year)	8	7
Production Rate (Units/Year)	665	875
Annual Shortage Cost (Rs./Unit)	90	110
Fraction of the Shortage Quantity Which Is Not Backlogged	0.25	0.2

Evaluate the optimum values of:

a. Common cycle time
b. Total cost
c. Lot size for each item
d. Shortage quantity for each item

7. Derive the following for multi-item production environment when there is different cycle time for different items without stock outs:

a. Annual carrying cost
b. Annual setup cost
c. Common cycle time

8. Various parameters for the two items are as follows:

	Item 1	Item 2
Setup Cost (Rs.)	120	140
Demand (Units/Year)	320	380
Carrying Cost (Rs./Unit-Year)	6	8
Production Rate (Units/Year)	700	800

Evaluate the optimum values of:

a. Cycle time for each item
b. Total cost
c. Lot size for each item

9. Allow shortages in the multi-item production environment with different cycle times and derive:

a. Different cycle time for different items
b. Total cost
c. Lot size for each item
d. Shortage quantity for each item

10. Various parameters for the two items are as follows:

	Item 1	Item 2
Setup Cost (Rs.)	120	140
Demand (Units/Year)	320	380
Carrying Cost (Rs./Unit-Year)	8	7
Production Rate (Units/Year)	650	850
Annual Shortage Cost (Rs./Unit)	50	80

Evaluate the optimum values of:

a. Cycle time for each item
b. Total cost
c. Lot size for each item
d. Shortage quantity for each item

11. What do you understand by the shelf life constraint? Explain how to deal with it implementing:

a. Common cycle time approach
b. Different cycle time approach

12. Various parameters for the two items are as follows:

	Item 1	Item 2
Setup Cost (Rs.)	120	150
Demand (Units/Year)	315	370
Carrying Cost (Rs./Unit-Year)	6	8
Production Rate (Units/Year)	680	810

Select a suitable shelf life for each item in a year so that the shelf life constraint can be demonstrated. To deal with the shelf life, apply the following approaches:

a. Common cycle time approach
b. Different cycle time approach

13. Include the input item procurement, along with the multiproduct manufacturing scenario and derive the optimal values of:

    a. Cycle time
    b. Number of orders of input item for each finished product
    c. Total cost

14. Evaluate the optimal values of:

    a. Cycle time
    b. Number of orders of input item for each finished product
    c. Total cost

    assuming the following data:

	Finished Item 1	Finished Item 2
Ordering Cost of Input Item Corresponding to Finished Item (Rs.)	25	35
Setup Cost for Finished Item (Rs.)	120	145
Annual Demand for Finished Item (Units/Year)	300	400
Holding Cost of Input Item Related to End Product (Rs./Unit-Year)	5	6
Carrying Cost for Finished Product (Rs./Unit-Year)	7	8
Production Rate for Finished Product (Units/Year)	700	850

15. Incorporate the generalized production cost of running a facility and derive the optimal values of:

    a. Cycle time
    b. Total cost

16. Let:

    Production time cost per year = Rs. 9,500
    Shop floor index = 0.3

Various parameters for the two items are as follows:

	Item 1	Item 2
Setup Cost (Rs.)	120	140
Demand (Units/Year)	320	380
Carrying Cost (Rs./Unit-Year)	6	8
Production Rate (Units/Year)	700	800

Evaluate the optimum values of:

a. Cycle time
b. Total cost

Also conduct the sensitivity analysis in respect of:

a. Shop floor index
b. Variation in production rate

7 Manufacturing Rate Flexibility

A flexible production-inventory system, along with an adjustable production rate, might be useful in the following situations, among others:

1. Change in a human resources position
2. Change in equipment utilization
3. Worker-equipment interaction
4. Demand fluctuation

These four situations are discussed briefly here:

1. **Change in a human resources position:** Human resources at various levels (such as workers, executives, engineers, managers) may fluctuate in terms of their numbers. This might be because some may join and some may resign, and therefore, a fair chance of increase/decrease in human resources positions always exists. Accordingly, significant parameters such as production rate/capacity are affected. The system has to adjust to such fluctuations.
2. **Change in equipment utilization:** Equipment may be obsolete or on the verge of getting scrapped. This affects production rate. Also because of major maintenance problems, this parameter is influenced considerably. A piece of equipment may not be capable enough to produce the desired quality level, and the rate of production is adversely affected. Since one or some of the available pieces of equipment are not used for such specific items, the manufacturing or production rate decreases concerning these. In the reverse scenario, an increase takes place when these are available for some items where the desired quality level is not that stringent.
3. **Worker-equipment interaction:** Because of a change in the availability of equipment and present workers, it may not be possible to maintain a similar production rate for long time. This may be one of the various reasons for contributing towards fluctuations in the manufacturing rate, upwards or downwards.
4. **Demand fluctuation:** Customer demands may fluctuate on the up side or on the down side, and therefore, flexibility in the production rate might be helpful for the management.

Production rates need to be adjusted as per the demand in case of the flexible system. Otherwise, demand may fluctuate, and therefore, it is of interest to study the demand variation in the present context.

DOI: 10.1201/9781003213994-7

7.1 DEMAND VARIATION

Demand may vary on the up side, as well as on the down side.

7.1.1 Upward Variation

Product demand may experience an upward variation because of the following:

1. There might be a genuine change in the consumption pattern. Growth may also be observed in the context of product life cycle.
2. Sometimes, an increase in the demand may be affected by reviving the consumer relationship. It may be without incurring any additional cost, i.e. just by telephonic discussion or similar.

An existing production-inventory situation is shown in Figure 7.1.

Let:

Annual demand = R
Demand rate per period = r
Production rate per period = p
Proportion of nondefective items in a lot = y
Ordering quantity = Q
Setup cost = C
Unit production cost = P
Annual inventory holding cost per unit = H

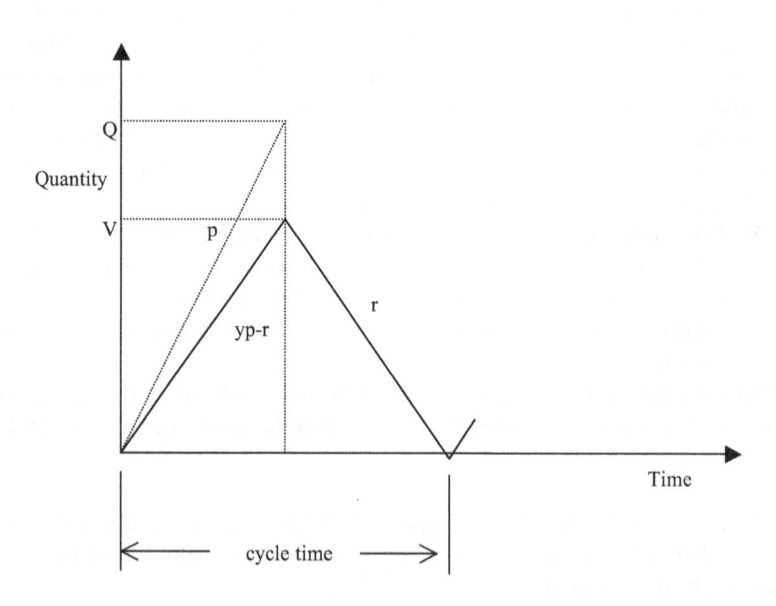

FIGURE 7.1 Existing production-inventory situation

Maximum inventory, $V = \dfrac{(yp - r)Q}{p}$

And annual inventory carrying cost $= \dfrac{V}{2} \cdot H$

$$= \frac{(yp - r)HQ}{2p}$$

Adding setup and production cost, total annual cost,

$$E = \frac{(yp - r)HQ}{2p} + \frac{RC}{yQ} + \frac{RP}{y} \tag{7.1}$$

Optimal production quantity,

$$Q^* = \sqrt{\frac{2pRC}{(yp - r)Hy}} \tag{7.2}$$

Substituting Q* in Equation 7.1, optimal total annual cost:

$$E^* = \frac{RP}{y} + \sqrt{\frac{2(yp - r)HRC}{yp}} \tag{7.3}$$

Upward variation is shown in Figure 7.2. Demand rate r is increased to r_1, and therefore inventory build-up rate corresponds to $(yp - r_1)$. This affects an annual demand also and R can be replaced by R_1 accordingly in the formulation.

Now the revised relevant cost:

$$E_1^* = \frac{R_1 P}{y} + \sqrt{\frac{2(yp - r_1)HR_1 C}{yp}} \tag{7.4}$$

It is if interest to determine when this demand increase also results into an overall cost reduction in addition to a potential profit increase. For this purpose,

$$E^* - E_1^* > 0.$$

Using Equation 7.3 and Equation 7.4, and after solving, it can be shown that:

$$C > \frac{yp}{2H}\left[\frac{P(R_1 - R)/y}{\sqrt{(yp - r)R} - \sqrt{(yp - r_1)R_1}}\right]^2 \tag{7.5}$$

If setup cost C is greater than the R.H.S. of Expression 7.5, then an overall cost reduction is certain. Such developed condition is useful for the companies to analyze their operational parameters pertaining to this activity and subsequent decision making.

Example 7.1

Consider:

Annual demand, R = 360 units
Setup cost, C = Rs. 900

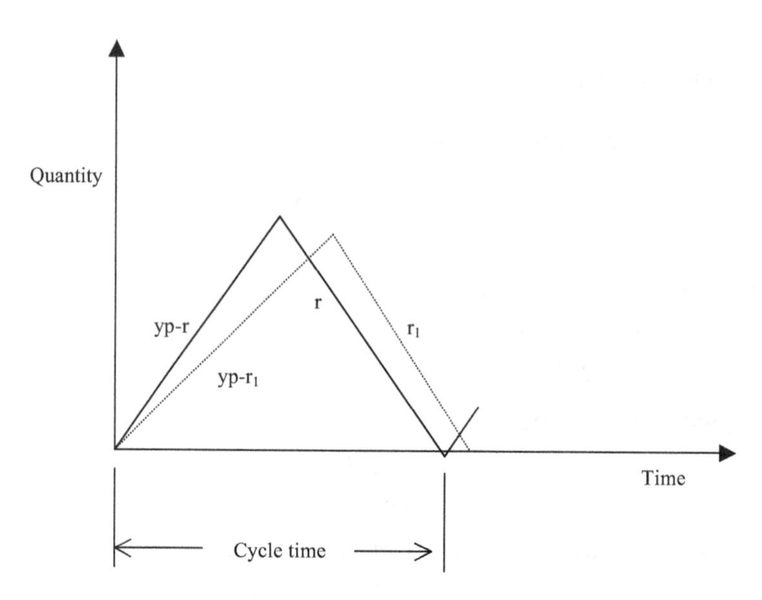

FIGURE 7.2 Upward demand variation

Proportion of nondefective items in a lot, y = 1.0
Annual inventory carrying cost per unit, H = Rs. 90
Production cost per unit, P = Rs. 10
Demand rate per period, r = 30 units
Production rate per period, p = 40 units
Increased demand rate per period, r_1 = 31 units
Increased annual demand, R_1 = 372 units

From Eq. (7.3), E* = Rs. 7,418.38.
From Eq. (7.4), E_1* = Rs. 7,402.31.

In spite of an increase in the demand, the total cost can decrease under certain circumstances. This is in addition to any profit gain concerning additional units.

7.1.2 Downward Variation

Demand may experience a downward variation because of the following:

1. Decrease in demand may be observed due to a genuine change in the pattern of consumer behaviour.
2. Due to a change in the business environment, the customer or company may cancel the order for a specific item, resulting in a demand decrease.

Downward variation is shown in Figure 7.3. Demand rate r is decreased to r_1, affecting the inventory build-up rate and the total relevant cost. Although downward

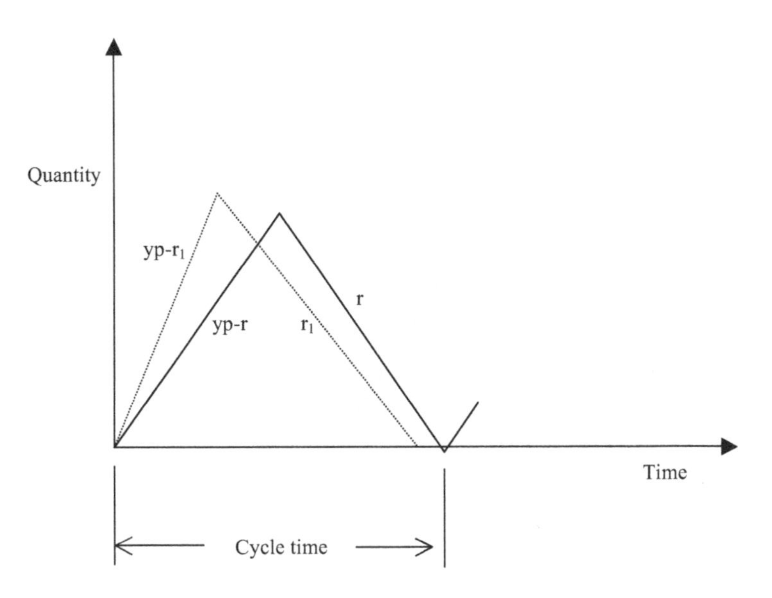

FIGURE 7.3 Downward demand variation

variation may not be preferred, it may be worth analyzing under such undesirable/unavoidable situations.

For certain cost reduction, it can be shown that:

$$C < \frac{yp}{2H}\left[\frac{P(R-R_1)/y}{\sqrt{(yp-r_1)R_1}-\sqrt{(yp-r)R}}\right]^2 \tag{7.6}$$

Example 7.2

Consider the base data of previous Example 7.1, except demand increase. Assume that:

Decreased demand rate per period, r_1 = 29 units
Decreased annual demand, R_1 = 348 units

From Equation 7.3, E* = Rs. 7,418.38.
From Equation. 7.4, E_1* = Rs. 7,417.43.

7.2 PRODUCTION RATE VARIATION

Production rate can be flexible enough so that it is possible to vary it. This can happen because of the following:

1. A machine may have a varied range for running, and therefore the manufacturing rate can be increased or decreased with reference to the existing one.
2. In certain cases, an overall rate of production can be varied by an increase or decrease of human resources on the production line.

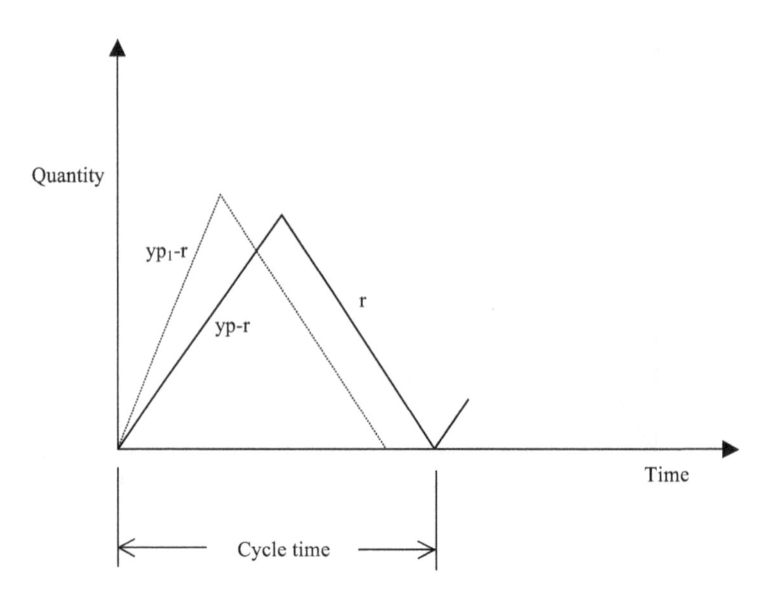

FIGURE 7.4 Upward production rate variation

For the production rate variation, the existing situation should be formulated in a different manner. This is because production cost depends on the production time spent in that activity. If the production time cost (in Rs. per unit period) is A, then the annual production cost is:

$$\frac{R}{y} \cdot \frac{A}{p} = \frac{RA}{yp}$$

$$\text{The total annual cost, } E = \frac{(yp - r)HQ}{2p} + \frac{RC}{yQ} + \frac{RA}{yp} \tag{7.7}$$

Following similar procedure,

$$E^* = \frac{RA}{yp} + \sqrt{\frac{2(yp - r)HRC}{yp}} \tag{7.8}$$

7.2.1 Upward Variation

Upward variation is shown in Figure 7.4. Production rate p is increased to p_1.

$$E_1^* = \frac{RA}{yp_1} + \sqrt{\frac{2(yp_1 - r)HRC}{yp_1}} \tag{7.9}$$

For certain cost reduction, (E*—E1*) > 0 shows:

$$C < \frac{y}{2HR} \left[\frac{(RA / y)\{(1/p) - (1/p_1)\}}{\sqrt{(yp_1 - r)/p_1} - \sqrt{(yp - r)/p}} \right]^2 \tag{7.10}$$

Example 7.3

Consider:

Annual demand, R = 360 units
Setup cost, C = Rs. 900
Proportion of nondefective items in a lot, y = 1.0
Annual inventory carrying cost per unit, H = Rs. 90
Production time cost per unit period, A = Rs. 1,000
Demand rate per period, r = 30 units
Production rate per period, p = 40 units
Increased production rate per period, p_1 = 41 units

From Equation 7.8, E* = Rs. 12,818.37
From Equation 7.9, E_1* = Rs. 12,736.09

With the increased production rate, an overall cost reduction is observed because the assumed setup cost satisfies the Expression 7.10. Such analysis helps the management in creating suitable decision support system where it becomes convenient to know the potential economic operational scenario.

7.2.2 Downward Variation

Downward variation is shown in Figure 7.5, where production rate has been decreased from p to p_1 along with the proportion of nondefective items in a lot.

Using Equation 7.8 and Equation 7.9, and following similar procedure, it can be shown that:

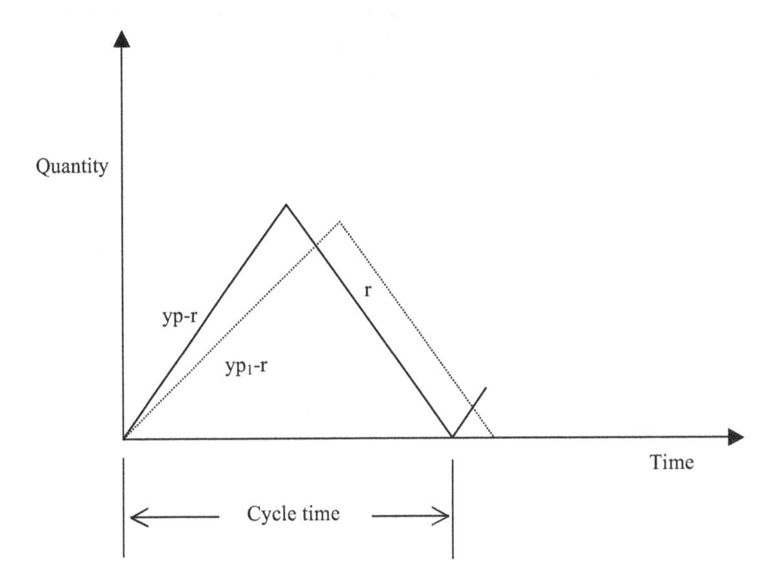

FIGURE 7.5 Downward production rate variation

$$C > \frac{y}{2HR}\left[\frac{(RA/y)\{(1/p_1)-(1/p)\}}{\sqrt{(yp-r)/p}-\sqrt{(yp_1-r)/p_1}}\right]^2 \tag{7.11}$$

Expression 7.11 will help in knowing when a decrease in production rate is economical.

Example 7.4

Consider the data of Example 7.3.

From Equation 7.11, C > 2,135.90.
Therefore, for demonstrating economical case, assume C = Rs. 2,150.

Additionally,

Decreased production rate per period, p_1 = 39 units.

Now total cost before and after the downward variation in production rate are observed as follows:

E* = Rs. 14,901.70
E1* = Rs. 14,900.93.

7.3 INCLUDING SHORTAGES

In some cases shortages might not be avoided. If shortage costs are estimated in the business environment, these can be included in the modelling process, and relevant costs are optimized along with the evaluation of suitable parameters. Inclusion of shortages can be visualized as shown in Figure 7.6.

Assume that all the shortages are completely backordered. Also:

J = maximum shortage quantity
K = annual shortage cost per unit.

Maximum inventory, $V = (yp-r)\dfrac{Q}{p} - J$

And annual inventory carrying cost $= \dfrac{pV^2H}{2Q(yp-r)}$

$$= \frac{(yp-r)HQ}{2p} - HJ + \frac{HpJ^2}{2(yp-r)Q}$$

Adding setup, production, and shortage cost also, total annual cost,

$$E = \frac{pJ^2(K+H)}{2(yp-r)Q} + \frac{(yp-r)HQ}{2p} - HJ + \frac{RP}{y} + \frac{RC}{yQ} \tag{7.12}$$

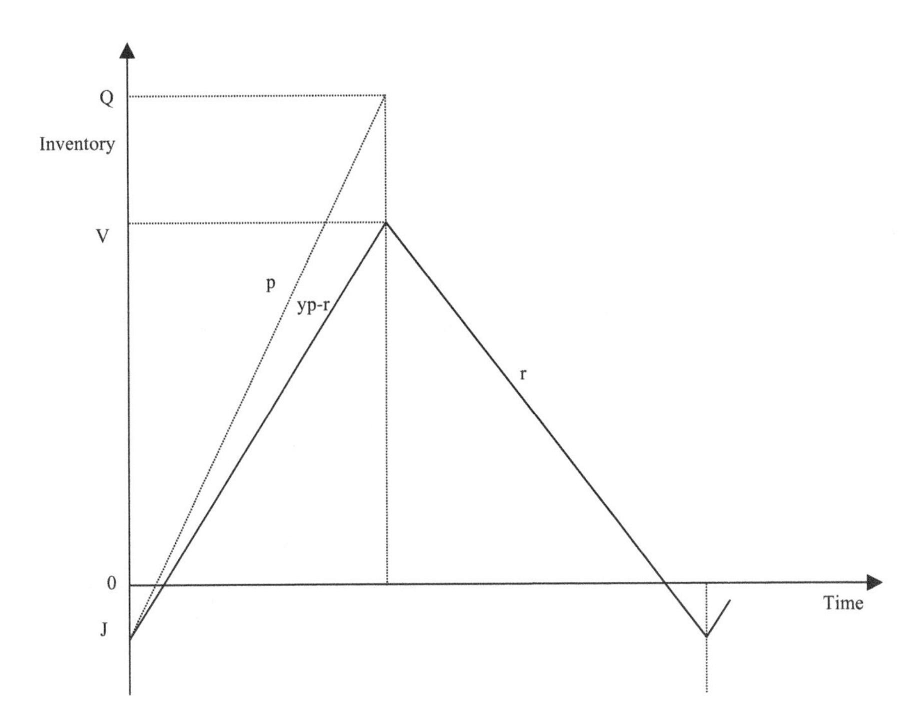

FIGURE 7.6 Existing situation with shortages

It can be shown that the optimal values are:

$$Q^* = \sqrt{\frac{2pRC(K+H)}{(yp-r)HyK}} \tag{7.13}$$

$$J^* = \sqrt{\frac{2RCH(yp-r)}{ypK(K+H)}} \tag{7.14}$$

$$E^* = \frac{RP}{y} + \sqrt{\frac{2RCHK(yp-r)}{yp(K+H)}} \tag{7.15}$$

7.3.1 Demand Variation

Upward demand variation is shown in Figure 7.7.
The revised cost:

$$E_1^* = \frac{R_1 P}{y} + \sqrt{\frac{2R_1 CHK(yp-r_1)}{yp(K+H)}} \tag{7.16}$$

E* — E1* > 0 shows:

$$C > \frac{yp(K+H)}{2HK}\left[\frac{P(R_1-R)/y}{\sqrt{R(yp-r)}-\sqrt{R_1(yp-r_1)}}\right]^2 \tag{7.17}$$

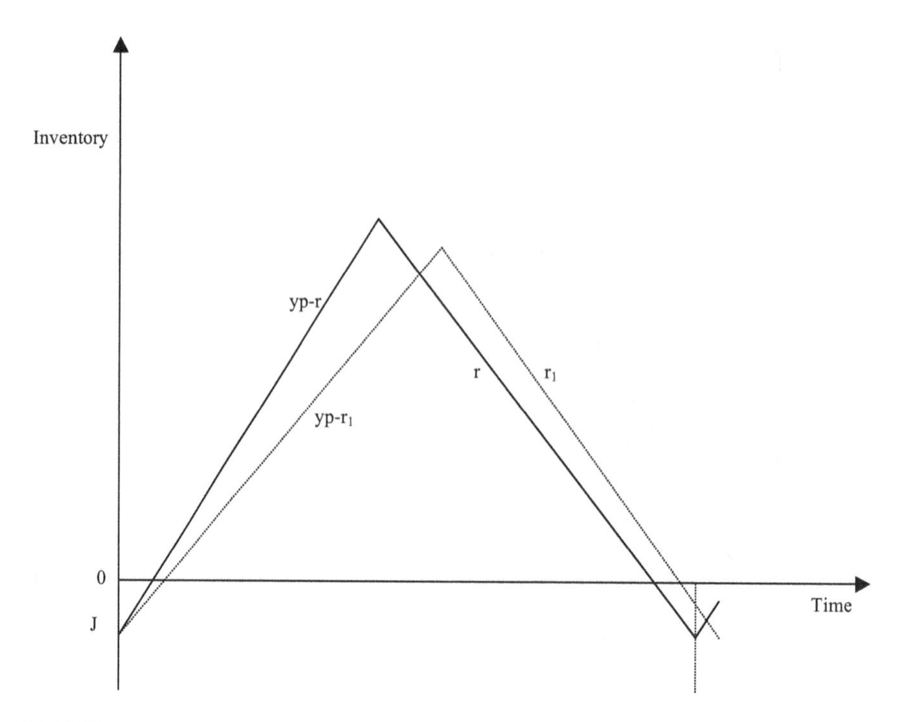

FIGURE 7.7 Upward demand variation with shortages

Example 7.5

Consider:

Annual demand, R = 360 units
Setup cost, C = Rs. 900
Proportion of nondefective items in a lot, y = 1.0
Annual inventory carrying cost per unit, H = Rs. 90
Production cost per unit, P = Rs. 10
Demand rate per period, r = 30 units
Production rate per period, p = 40 units
Annual shortage cost per unit, K = Rs. 500
Increased demand rate per period, r_1 = 31 units
Increased annual demand, R_1 = 372 units

From Equation 7.15, E* = Rs. 7,115.10.
From Equation 7.16, E_1* = Rs. 7,109.84.

Downward demand variation is shown in Figure 7.8.
For this situation, it can be shown that:

$$C < \frac{yp(K+H)}{2HK}\left[\frac{P(R-R_1)/y}{\sqrt{(yp-r_1)R_1}-\sqrt{(yp-r)R}}\right]^2 \tag{7.18}$$

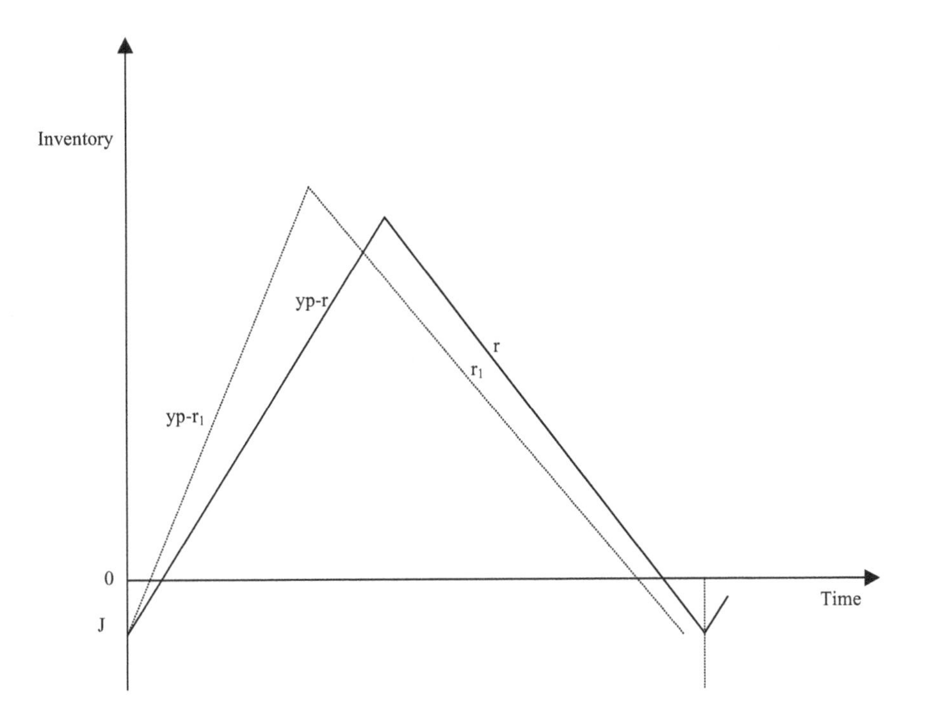

FIGURE 7.8 Downward demand variation with shortages.

Example 7.6

Consider the base data of previous Example 7.5, except demand increase. Assume that:

Decreased demand rate per period, $r_1 = 29$ units
Decreased annual demand, $R_1 = 348$ units

Now, the reduced cost, $E_1^* =$ Rs. 7,104.70.

7.3.2 Production Rate Variation

The upward production rate variation is represented by Figure 7.9, along with shortages. As discussed before, assume:

Production time cost per unit period = A

And the relevant optimal cost:

$$E^* = \frac{RA}{yp} + \sqrt{\frac{2RCHK(yp - r)}{yp(K + H)}} \tag{7.19}$$

After the production rate increase,

$$E_1^* = \frac{RA}{yp_1} + \sqrt{\frac{2RCHK(yp_1 - r)}{yp_1(K + H)}} \tag{7.20}$$

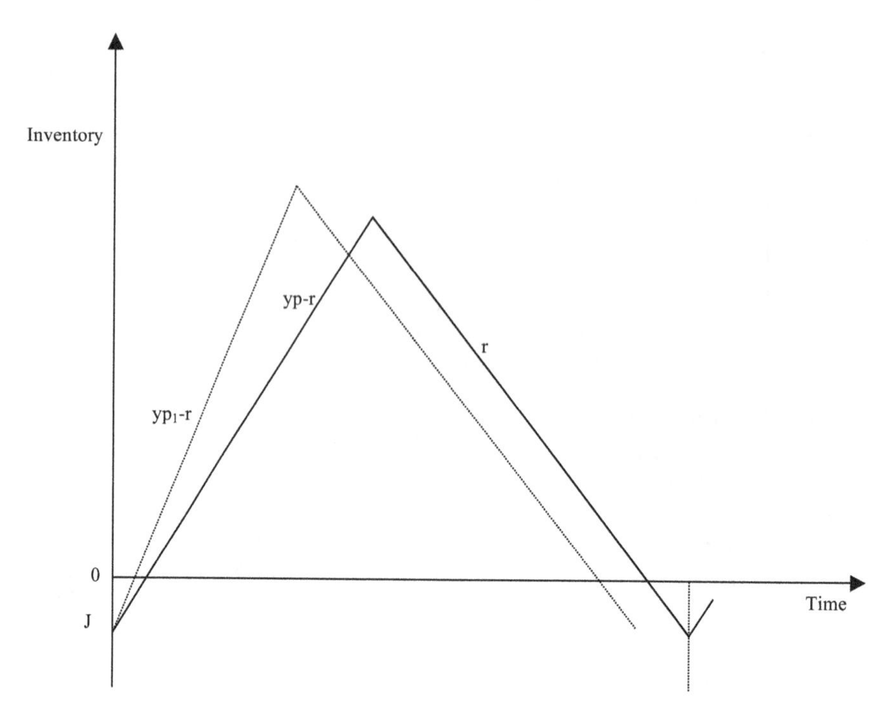

FIGURE 7.9 Upward production rate variation with shortages

For an economical case, it can be derived that:

$$C < \frac{y(K+H)}{2HRK}\left[\frac{(RA/y)\{(1/p)-(1/p_1)\}}{\sqrt{(yp_1-r)/p_1}-\sqrt{(yp-r)/p}}\right]^2 \tag{7.21}$$

Example 7.7

Consider:

Annual demand, R = 360 units
Setup cost, C = Rs. 900
Proportion of nondefective items in a lot, y = 1.0
Annual inventory carrying cost per unit, H = Rs. 90
Production time cost per unit period, A = Rs. 1,000
Demand rate per period, r = 30 units
Production rate per period, p = 40 units
Increased production rate per period, p_1 = 41 units
Annual shortage cost per unit, K = Rs. 500

From Equation 7.8, E* = Rs. 12,818.37
From Equation 7.9, E_1* = Rs. 12,736.09

Figure 7.10 shows the downward production rate variation.

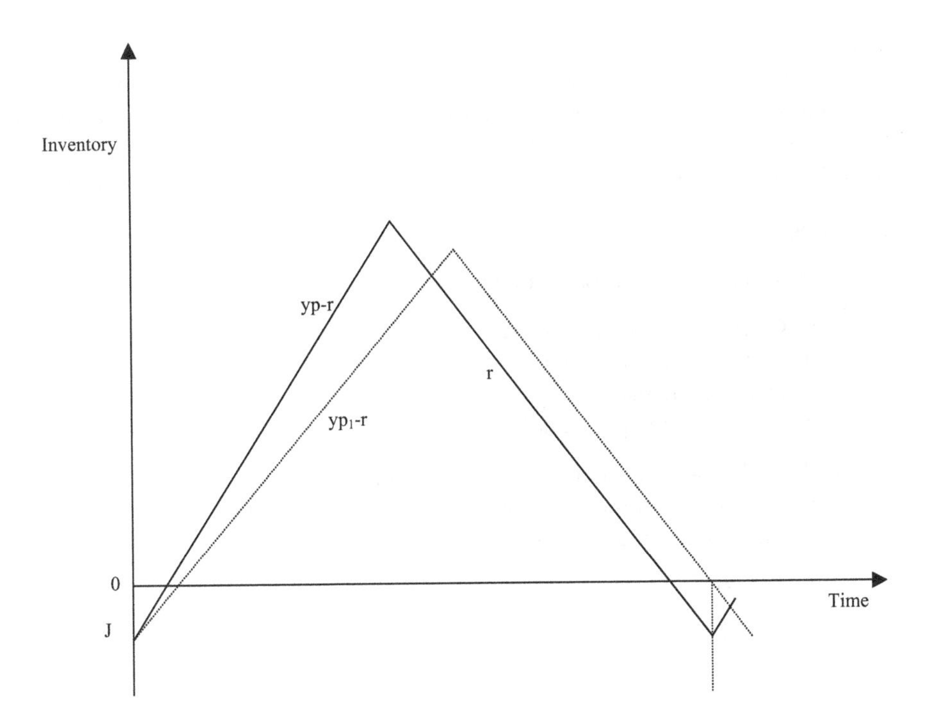

FIGURE 7.10 Downward production rate variation with shortages

It can be derived that:

$$C > \frac{y(K+H)}{2HRK}\left[\frac{(RA/y)\{(1/p_1)-(1/p)\}}{\sqrt{(yp-r)/p}-\sqrt{(yp_1-r)/p_1}}\right]^2 \tag{7.22}$$

Example 7.8

Consider:

Annual demand, R = 360 units
Setup cost, C = Rs. 3,000
Proportion of nondefective items in a lot, y = 1.0
Annual inventory carrying cost per unit, H = Rs. 90
Production time cost per unit period, A = Rs. 1,000
Demand rate per period, r = 30 units
Production rate per period, p = 40 units
Decreased production rate per period, p_1 = 39 units
Annual shortage cost per unit, K = Rs. 500

Now the following cost values are obtained:

E* = Rs. 15,417.67
E_1* = Rs. 15,396.66

With the proposed formulations, it is possible to analyze the flexibility in production rate. Depending on the demand variation, as well as the resource variation or other problems, production rate needs to be varied. The conditions concerning the setup cost have been developed. There is certainty of cost reduction if such condition is satisfied in the business environment, as per the respective data set or operational parameters.

Exercises

1. In the context of a production-inventory system, discuss the following:
    a. Change in a human resources position
    b. Worker-equipment interaction
    c. Demand fluctuation
    d. Change in equipment utilization
2. What do you understand by the demand rate variation?
   In case of the infinite replenishment rate as shown in the following figure, analyze the upward as well as downward variation.

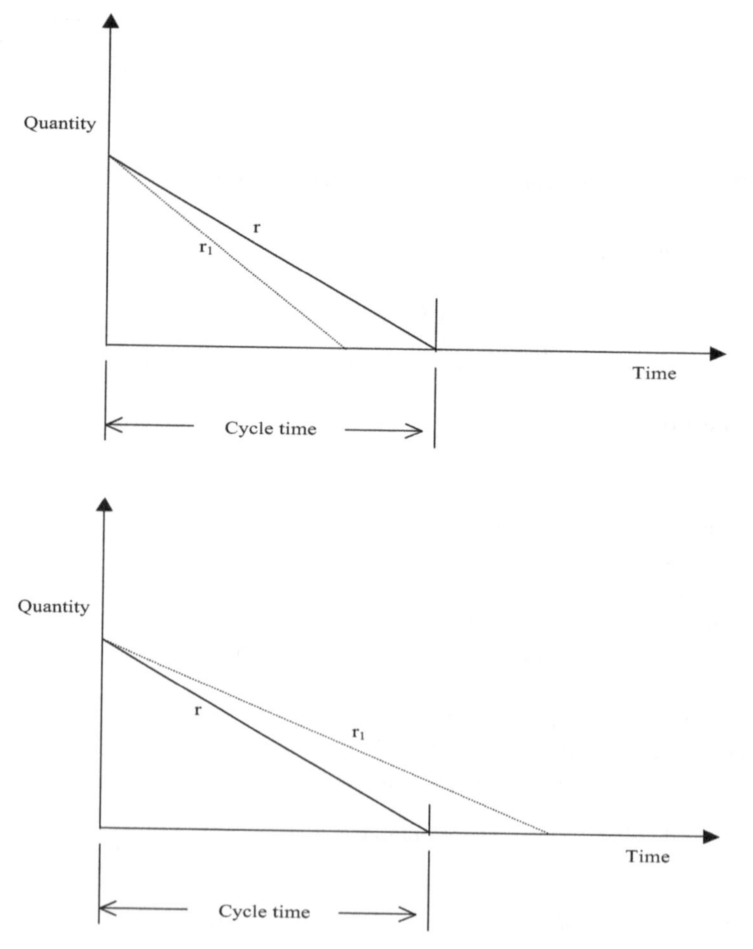

3. Consider the following data:

   Annual demand = 360 units
   Ordering cost = Rs. 100
   Proportion of nondefective items in a lot = 0.9
   Annual inventory carrying cost per unit, H = Rs. 3
   Purchase cost per unit, P = Rs. 10
   Demand rate per period, r = 30 units

   Comment on the result if demand rate per period increases or decreases by 10%.
4. If shortages are incorporated as shown in the following figure, analyze the upward variation, as well as downward variation, in demand rate.

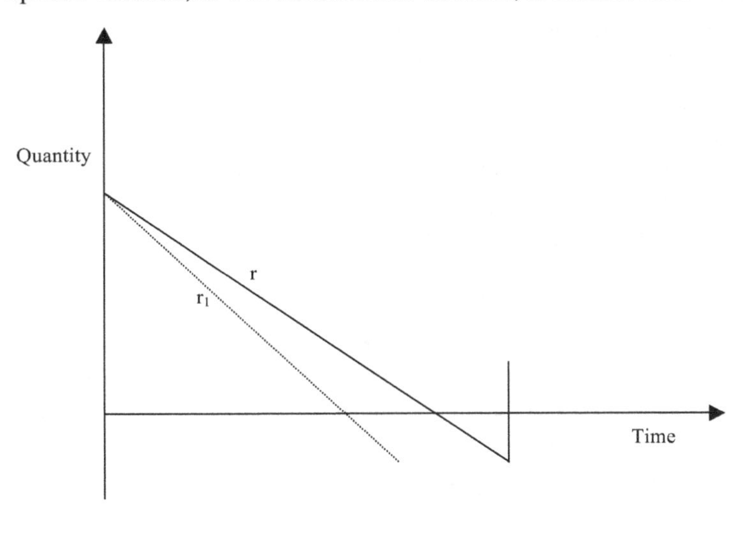

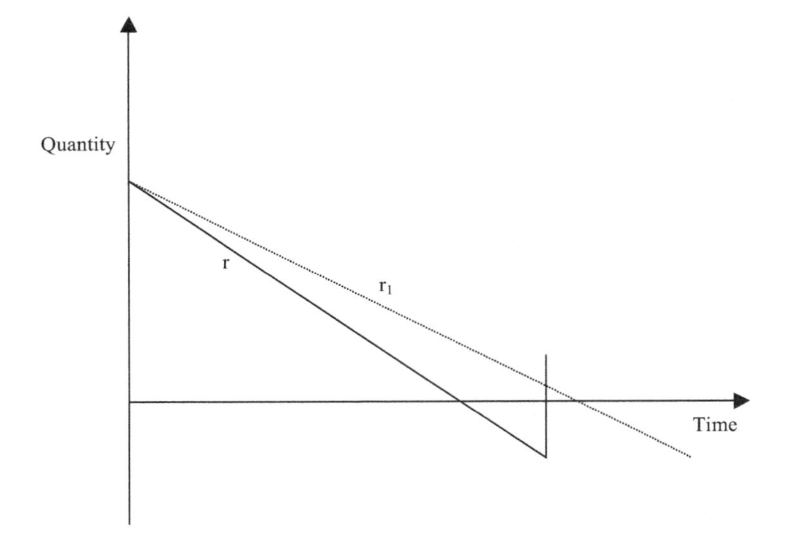

5. Consider the following data:

   Annual demand = 360 units
   Ordering cost = Rs. 100
   Proportion of nondefective items in a lot = 0.9
   Annual inventory carrying cost per unit, H = Rs. 3
   Purchase cost per unit, P = Rs. 10
   Demand rate per period, r = 30 units
   Annual shortage cost per unit = Rs. 50

   Comment on the result if demand rate per period increases or decreases by 10%.
6. In a production setup, explain the upward variation in the demand. Derive the necessary condition which will ensure the cost reduction.
7. Consider the following parameters:

   Annual demand = 300 units
   Setup cost = Rs. 800
   Proportion of nondefective items in a lot = 0.95
   Annual inventory carrying cost per unit = Rs. 70
   Production cost per unit = Rs. 20
   Demand rate per period = 25 units
   Production rate per period = 35 units

   Comment on the result if demand rate per period increases by two units.
8. In a production setup, explain the downward variation in the demand. Derive the necessary condition which will guarantee the cost reduction.
9. Consider the following parameters:

   Annual demand = 360 units
   Setup cost = Rs. 700
   Proportion of nondefective items in a lot = 0.99
   Annual inventory carrying cost per unit = Rs. 75
   Production cost per unit = Rs. 15
   Demand rate per period = 30 units
   Production rate per period = 40 units

   Comment on the result if demand rate per period decreases by two units.
10. In a batch manufacturing environment, discuss the effect of an increase in manufacturing rate. In order to ascertain economical manufacturing rate variation, develop a suitable expression. Explain the role of manufacturing setup cost in such environment.
11. Calculate and compare the relevant costs if production rate per period increases by three units, assuming the input data as follows:

    Annual demand = 300 units
    Setup cost = Rs. 600
    Proportion of nondefective items in a lot = 0.98
    Annual inventory carrying cost per unit = Rs. 70

Production time cost per unit period = Rs. 950
Demand rate per period = 25 units
Production rate per period = 35 units

12. In a batch manufacturing environment, discuss the effect of a decrease in manufacturing rate. In order to ascertain economical manufacturing rate variation, develop a suitable expression. Explain the role of manufacturing setup cost in such environment.
13. Calculate and compare the relevant costs if production rate per period decreases by one unit, considering the input parameters as follows:

    Annual demand = 360 units
    Setup cost = Rs. 2,300
    Proportion of nondefective items in a lot = 0.99
    Annual inventory carrying cost per unit = Rs. 85
    Production time cost per unit period = Rs. 1,000
    Demand rate per period = 25 units
    Production rate per period = 40 units

14. What do you understand to be meant by the shortages? After an inclusion of shortages, obtain the optimal values of the following output parameters:

    a. Batch quantity
    b. Shortage quantity
    c. Total relevant cost

15. Include the shortages/shortage cost in the batch production-inventory analysis for the following scenario:

    a. Upward demand variation
    b. Downward demand variation
    c Upward production rate variation
    d. Downward production rate variation

16. Use the following parameters:

    Proportion of nondefective items in a lot, y = 1.0
    Annual inventory carrying cost per unit, H = Rs. 95
    Production cost per unit, P = Rs. 12
    Demand rate per period, r = 32 units
    Production rate per period, p = 41 units
    Annual shortage cost per unit, K = Rs. 550

    Find out the desired upper and lower bounds for the production setup cost in order to ensure cost improvement with the demand rate variation of two units on both sides. Also evaluate the costs for the following situations:

    a. Increased demand rate
    b. Decreased demand rate

17. The given input information is as follows:

    Annual inventory carrying cost per unit, H = Rs. 90
    Annual shortage cost per unit, K = Rs. 450
    Production time cost per unit period, A = Rs. 1,100
    Demand rate per period, r = 28 units
    Proportion of nondefective items in a lot, y = 0.99
    Production rate per period, p = 38 units

    Explain the approach to obtain suitable value for the manufacturing setup cost in order to make the production rate variation economical for the cases:

    a. Downward production rate variation
    b. Upward production rate variation

    For the stated purpose, production rate per period can be assumed to vary by three units on both sides, i.e. downward and upward.

Additional Reading

Manufacturing inventory and its flow also depends on the material flow lines and the layout of facilities/machines. The present discussion includes material flow pattern and the layout planning procedures, among the other related aspects.

The raw material/input items move from one machine to another machine until these become the finished item. The material flow lines refer to the flow pattern. There are five basic flow lines:

1. I-Flow
2. L-Flow
3. S-Flow
4. O-Flow
5. U-Flow

The flow lines are according to the shape of the letter, i.e., 'I', 'L', 'S', 'O', or 'U'.

1. I-Flow

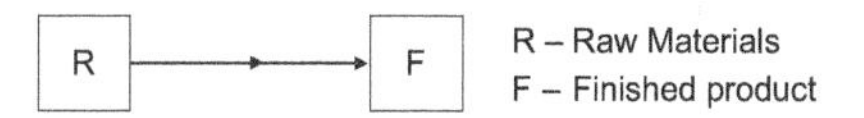

At location 'R', the raw material enters and at location 'F', the finished product is available for dispatch. The machines and facilities are arranged to the line joining 'R' and 'F' as per the requirement. The type of flow line is suitable for the available space, which is more in length and less in width.

2. L-Flow

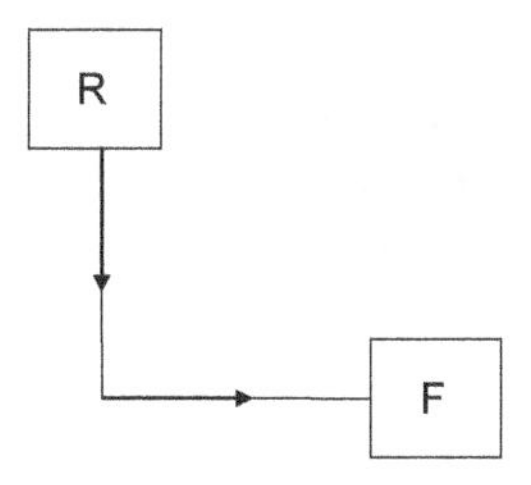

L-Flow is used to take an advantage of width of the plot also, if it is sufficiently wide.

3. S-Flow

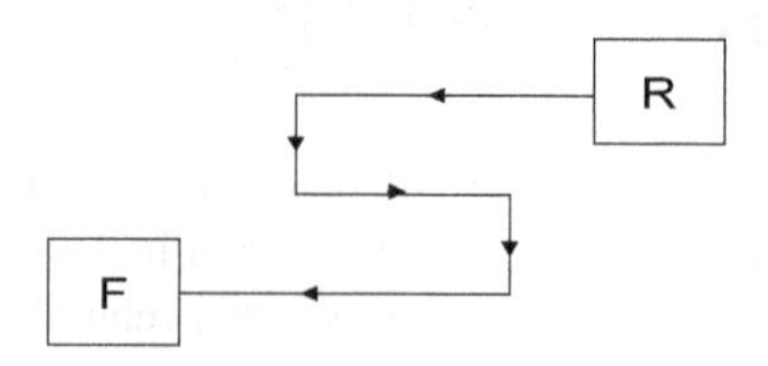

When the number of facilities are more and the available space suits, S-Flow may be adopted. It can also be imagined as a combination of U-Flow and L-Flow lines, and some of the advantages of both the patterns can be availed.

4. O-Flow

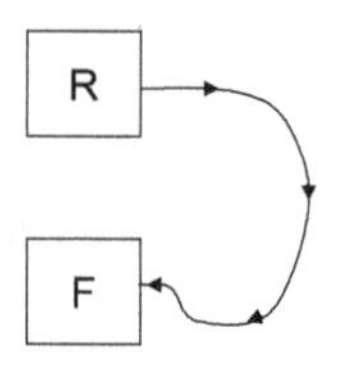

5. U-Flow

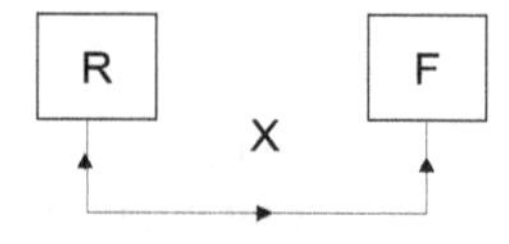

The facilities are arranged in U-shape. The engineers and supervisors need to travel less distance. For example, it may be convenient to supervise the machines surrounding a person standing at location 'X'.

Out of the five basic flow lines, in case of O-Flow and U-Flow, the entry of raw material/input items 'R' and dispatch of finished products 'F' are on the same side of the plant. Vehicles will be coming in and going out for the transportation of raw materials and finished items, respectively. Road is needed to be built up on only one side of the plant. But at the same time, a heavy rush of transportation will be there. If the final product is smaller in size, the boxes/bins/containers carrying them flow through the facilities. After dispatch of the final products, the empty boxes are needed to be brought back to the input side of the plant. It becomes convenient in O-Flow and U-Flow pattern because 'R' and 'F' are nearer to each other.

For I-Flow, L-Flow and S-Flow, the roads are required on the both sides of plant. But a heavy rush of transportation is not there because the activities are divided on two sides of the layout for transportation of input and finished items. The handling cost for empty containers may be high.

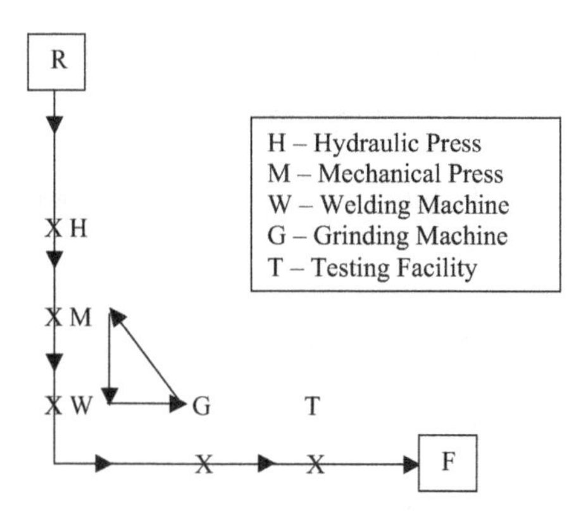

FIGURE 1 Repeated flow.

In an industry, there may be a combination of two or more types of basic flow lines.

Sometimes the repeated type of flow may also take place, as shown in Figure 1.

After the grinding operation is over, the part/product may again go to the mechanical press for another operation which is subsequent to first grinding job. This is known as the repeated flow. In the absence of repeated flow, the material flow is unidirectional.

In addition to these factors, vertical flow lines are adopted to take advantage of gravity. These may be suitable for chemical processing in which large vessels are placed on different floors, one over the other. The vertical flow lines are also useful when horizontal space constraints exist.

1 ASSEMBLY LINE BALANCING

Assembly line balancing is an important area of the analysis of product layout. Apart from other issues, it has implications for layout, particularly with reference to the workstation size. A work station is a physical location where one or more tasks are performed. The objective is to assign tasks to workstations in such a way that each workstation completes the assigned tasks in less than or equal to the cycle time. A cycle time is the time between successive products coming off the end of an assembly line.

$$\text{Cycle Time, C.T.} = \frac{\text{Production time per period}}{\text{Required production quantity per period}}$$

1.1 Largest Number of Following Tasks Rule

The requirements for implementing this rule are as follows:

1. Precedence diagram which is a representation of relationship among tasks. For example, given the information:

Task	Task Which Immediately Precedes
A	–
B	A
C	A
D	B,C
E	D

The precedence diagram can be drawn as follows:

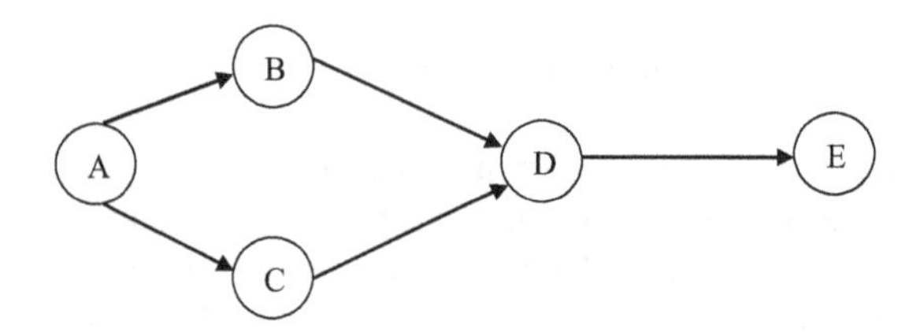

2. The required production quantity per period.
3. The time taken by each task, t_i
4. As the name of the rule suggests, assignment of tasks is in the order of the largest number of following tasks. It will be helpful if the number of tasks which follow each task is known. Consider the discussed precedence diagram. First task is A and four tasks follow it, viz. B, C, D and E. Two tasks, i.e. D and E, follow both B and C. Only one task E follows D, and no task follows E because E is the end task. This can be summarized as follows:

Task	Number of Tasks Which Follow
A	4
B, C	2
D	1
E	0

Example 1

An assembly line involves the tasks from A to I. A total of 360 products are required per day. Production time is eight hours per day. The time taken by each task and the tasks which immediately precede each task are:

Task	Tasks Which Immediately Precede	Time to Complete Task (Seconds)
A	–	40
B	A	30
C	A	35
D	C	30
E	C	20
F	B	45
G	D, E	50
H	F	30
I	G, H	35

The precedence relationship is shown in Figure 2.

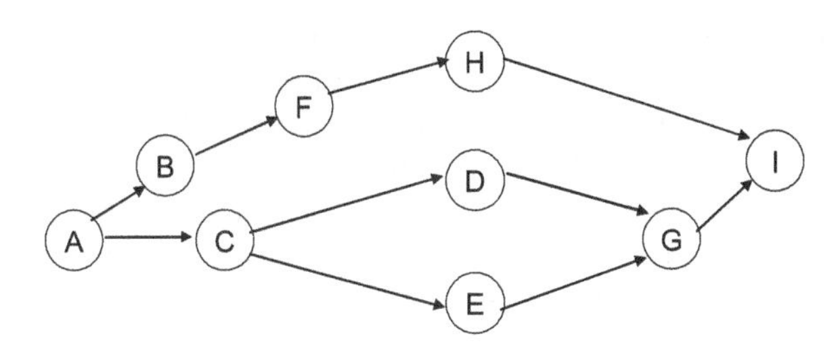

FIGURE 2 Precedence relationship.

As discussed before, it is necessary to know the total number of tasks which follow each task, because the assignment of task to each workstation will be in the order of largest number of following tasks. It is obtained as follows:

Task	A	B	C	D	E	F	G	H	I
Total Number of Tasks Which Follow It	8	3	4	2	2	2	1	1	0

Table 1 may be arranged in a decreasing order of the number of following tasks.

TABLE 1
Tasks in the Order of Largest Number of Following Tasks First

Task	A	C	B	D	E	F	G	H	I
Total Number of Tasks Which Follow It	8	4	3	2	2	2	1	1	0

Before assigning the tasks to workstation, the cycle time needs to be computed, because each workstation should not consume more than the cycle time in order to achieve the production target.

$$\text{Cycle Time} = \frac{\text{Production time per period}}{\text{Required production quantity per period}}$$
$$= \frac{8 \times 60 \times 60}{360}$$
$$= 80 \text{ seconds}$$

Refer to Table 1. To begin with, the first task in the order is A since it has the maximum number of following tasks.

Workstation 1

The first task that is assigned to Workstation 1 is A, which takes 40 seconds.

Time which is yet to be assigned = 80 – 40 = 40 seconds
Set of tasks which may be assigned at this stage considering precedence diagram (Figure 2) only, i.e., without checking the feasibility, S = [B, C].
Task time of B and C are 30 and 35 seconds respectively, which is less than the time yet to be assigned i.e. 40 seconds as calculated before. Set of feasible tasks which may be added to Workstation 1, T = [B, C].
From the set T, any one task will be selected based on the order of largest number of following tasks. From Table 1, C is to be given priority over B. Hence, C is selected for an addition to Workstation 1.

Remaining unassigned time = 40 – 35 = 5 seconds.

From Figure 2, once C is assigned, the tasks which follow it are D and E. At the same time, task B may also be considered because there is no such constraint, as task B cannot be performed after C.

Set of tasks which may be assigned (without checking the feasibility), S = [B, D, E]

Unassigned time is only 5five seconds, whereas B, D and E each take more than that.

Set of feasible tasks, T = [Nil]

Workstation 1 cannot accommodate any task further. This includes task A and C.

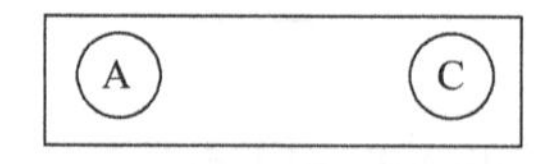

Workstation 1

Workstation 2

Set of tasks, S which was recently obtained = [B, D, E]. From this set, the first task for Workstation 2 is B because it is prior to D and E, as can be observed from Table 1.
Workstation 2 can also accommodate tasks whose total time does not exceed the cycle time, i.e. 80 seconds.
Time which is yet to be assigned = 80 – 30 = 50 seconds.

Now S = [D, E, F].

Set of feasible tasks, T = [D, E, F].

From Table 1, D, E and F have similar number of following tasks, i.e., two. There is a tie in applying the largest number of following tasks rule. This is resolved by selecting the task with the maximum operation time. Task time for D, E and F are 30, 20, and 45 seconds, respectively. As the task F takes maximum time, it is to be selected. (Note: In case of a tie further, any one task may be selected arbitrarily.)

Now remaining time = 50 – 45 = 5 seconds.

S = [D, E, H] and as each task takes more than five seconds,

T = [Nil]

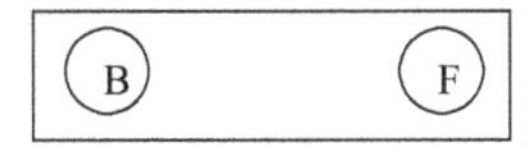

Workstation 2

Workstation 3

Recently obtained, S = [D, E, H].

First task to be selected for Workstation 3 is based on the largest number of following tasks. But there is a tie between D and E. Task time for D, i.e. 30 seconds, is more than that of E; therefore, D is selected.

Remaining time = 80 – 30 = 50 seconds

Again S = [E, H].

(Note: G is succeeding D, but it cannot be considered unless E is over. As D and E both are predecessors to G, G can be taken into consideration only when D and E both are complete.) (Refer to Figure 2.)

T = [E, H].

E can easily be selected from Table 1, as discussed before.

Remaining time = 50 – 20 = 30 seconds.

Now S = [G, H].

and T = [H] as G takes 50 seconds, which is more than the remaining time (30 seconds).

H is included in the workstation and as the remaining time = 30 – 30 = 0 seconds; no other task can be added.

However, the set of tasks without checking feasibility, S = [G].

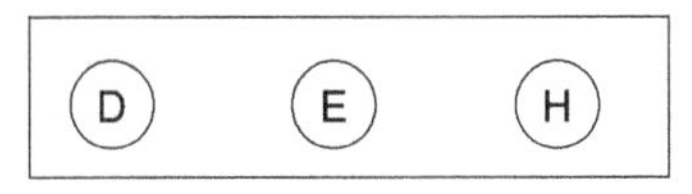

Workstation 3

Workstation 4

G is the first task for Workstation 4.

Remaining time = 80 – 50 = 30 seconds

S = [I], but T = [Nil]

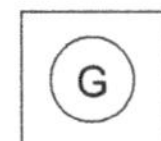

Workstation 4

Workstation 5

S = [I]

Only one task, i.e. the end task, remains to be included in this workstation.

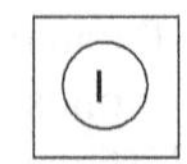

Workstation 5

In this way, all the tasks have been accommodated in the five workstations. Figure 3 shows the workstations with one or more tasks. Time in seconds for all the tasks is also shown.

In order to evaluate the performance of the proposed solution, i.e. a creation of five workstations, the efficiency needs to be calculated.

$$\text{Efficiency (\%)} = \frac{\text{Theoretical minimum number of workstations x 100}}{\text{actual number of workstations obtained}}$$

Here, actual number of workstations obtained = 5

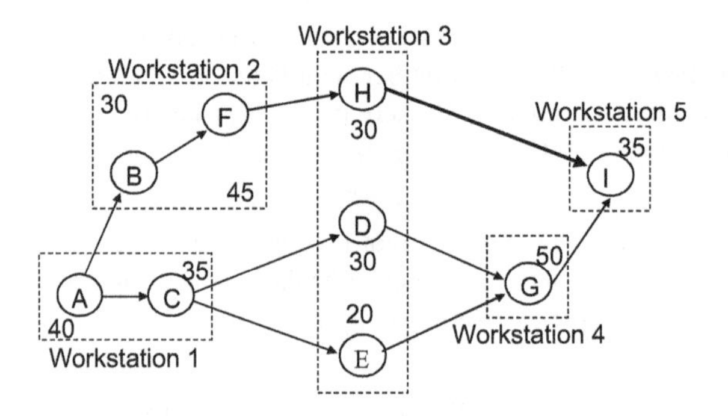

FIGURE 3 Workstations including one or more tasks.

and the theoretical minimum number of workstations

$$= \frac{\text{Total task time } (\Sigma t_i)}{\text{Cycle time}}$$

$$= \frac{40+30+35 + 30 + 20 + 45 + 50 + 30 + 35}{80}$$

$$= \frac{315}{80}$$

$$= 3.9375$$

$$\text{and efficiency of the proposal} = \frac{3.9375}{5} \times 100$$

$$= 78.75\%$$

The solution process is briefly explained in Table 2.

Summary of the Largest Number of Following Tasks Rule

1. Selection of the task, whether it is the first task in the workstation or subsequent tasks, will be made as follows:

   a. Task with largest number of following tasks is selected.
   b. If there is a tie in a., task with maximum operation time is selected.
   c. If there is a tie in b., any task is selected arbitrarily.

TABLE 2
Assembly Line Balancing Using the Largest Number of Following Tasks Rule

	Task	Time	Time Yet to be Assigned	Set of Tasks, S	Set of Feasible Tasks, T	Task with Largest Number of Followers	Task with Maximum Operation Time	Selected First Task for the Next Workstation
Workstation 1	A	40	40	[B,C]	[B,C]	C	—	—
	C	35	5	[B, D, E]	[Nil]	—	—	B
Workstation 2	B	30	50	[D,E,F]	[D,E,F]	D,E,F,	F	—
	F	45	5	[D,E,H]	[Nil]	—	—	D
Workstation 3	D	30	50	[E,H]	[E,H]	E	—	—
	E	20	30	[G,H]	[H]	H	—	—
	H	30	0	[G]	[Nil]	—	—	G
Workstation 4	G	50	30	[I]	[Nil]	—	—	I
Workstation 5	I	35	45	—	—	—	—	—

2. Workstation is a physical location where one or more tasks are performed. Workstation should be able to complete the assigned tasks at the most in a cycle time where,

$$\text{Cycle time} = \frac{\text{Production time per period}}{\text{Required production quantity per period}}$$

3. Precedence diagram is constructed.
4. The rule is implemented to balance the assembly line in the following steps:

   a. Select the first task as explained in 1. of this list, and assign it to Workstation 1. Compute the time yet to be assigned.
   b. Find the set of tasks S, considering precedence diagram only, i.e., without checking the feasibility with respect to time. From S, obtain the set of feasible tasks T by comparing each task time with the unassigned time. Select a task from T, as discussed in 1.
   c. If the set of feasible tasks T becomes empty, then no more tasks can be accommodated in the current workstation. Select one task from the recently obtained set S, as described in 1. Selected task becomes the first assigned task for the next workstation. Compute the time yet to be assigned.
   d. Repeat b. and c. until all the tasks are assigned.

5. To evaluate the performance of the proposed solution,

Efficiency (%)

$$= \frac{\text{Theoretical minimum number of workstations}}{\text{Actual number of workstations obtained in (4)}} \times 100$$

Whereas theoretical number of workstations,

$$= \frac{\text{Total task time } \left(\sum ti\right)}{\text{Cycle time}}$$

and t_i = time taken by task i.

Another heuristic to balance the assembly line is the longest operation time heuristic. In this rule, a task is selected primarily on the basis of the longest operation time first. In case of a tie, the task with the largest number of followers is selected.

1.2 Incremental Utilization Heuristic

In this procedure, the worker utilization is considered. The tasks are simply added one at a time to the workstation until the utilization of workers is 100% or it begins to decrease.

It is assumed that each worker in the group employed at a workstation does all the tasks assigned to that workstation. For example, the total task time for a workstation is 0.3 minutes (let us say, two tasks A and B are assigned to the workstation with the task time 0.16 and 0.14 minutes, respectively. Thus total task time is 0.16 + 0.14 = 0.3 minutes). If the cycle time is 0.1 minutes, then the number of workers required at the workstation would be 0.3/0.1 = 3 and the worker utilization would be 100%.

Theoretical minimum number of workers for a workstation

$$= \frac{\text{Total task time for workstation}}{\text{Cycle time}}$$

If a third task C with the time 0.14 minutes is added to the workstation, then the total task time for it will be 0.16 + 0.14 + 0.14 = 0.44 minutes. The theoretical minimum number of workers required for the workstation would be 0.44/0.1 = 4.4, but the actual number of workers will be integer, i.e. five.

Worker Utilization (%)

$$= \frac{\text{Theoretical minimum number of workers}}{\text{Actual number of workers}} \times 100$$

$$= \frac{4.4}{5} \times 100$$

$$= 88\%$$

Tasks are added to the workstation as per the precedence diagram. With the addition of each task, the theoretical and actual number of workers—and subsequently, worker utilization—is computed. The tasks are added to the workstation until utilization of workers is 100% or the worker utilization decreases. If the utilization of workers is decreased by an addition of task, then the particular task is excluded from the workstation and it becomes the first task to be included in the next workstation.

Example 2

An assembly line operates for 420 minutes per day, and 2,100 products are required to be assembled each day. The time taken by each task and the tasks which immediately precede each task are given below:

Tasks	Task Which Immediately Precede	Time Taken (Minutes)
A	–	0.63
B	A	0.17
C	B	0.82
D	C	0.91
E	C	0.16
F	D,E	0.33
G	F	0.66
H	G	0.21

1. Draw the precedence diagram.
2. Calculate the cycle time.
3. Calculate the theoretical minimum number of workers for the whole assembly line.
4. In order to minimize the workers' idle time, assign tasks to the workstations using the incremental utilization heuristic.
5. Evaluate the proposed solution by computing efficiency of the proposal.

Now,

1. The precedence diagram is as follows:

2. Cycle time $= \dfrac{\text{Production time per day}}{\text{Required production quantity per day}}$

$$= \frac{420}{2100}$$

$$= 0.2 \text{ min.}$$

3. Theoretical minimum number of workers for the whole assembly line,

$$= \frac{\text{Total task time for the complete assembly}}{\text{Cycle time}}$$

$$= \frac{0.63 + 0.17 + 0.82 + 0.91 + 0.16 + 0.33 + 0.66 + 0.21}{0.2}$$

$$= \frac{3.89}{0.2}$$

$$= 19.45$$

4. Table 3 shows the computation for an assembly line balancing using the incremental utilization heuristic.

First task in the precedence diagram is A, which is included in Workstation 1. Task time of A = 0.63 minutes.

If Workstation 1 comprises only one task, A, then the theoretical number of workers required for this workstation is,

$$= \frac{0.63}{\text{Cycle time}}$$

$$= \frac{0.63}{0.2}$$

$$= 3.15$$

As the minimum number is 3.15, the actual number of workers for Workstation 1 (if it has only task A) = 4 and the worker utilization is,

$$= \frac{3.15}{4}$$

$$= 0.7875 \text{ or } 78.75\%$$

TABLE 3
Assembly Line Balancing by the Incremental Utilization Heuristic

Task	A	A+B	C	C + D	C + D + E	E	E + F	E + F + G	E + F + G + H
Total Task Time (a)	0.63	.63 + .17 = 0.8	.82	.82 + 91 = 1.73	1.73 + .16 = 1.89	.16	.16 + .33 = 0.49	.49 + .66 = 1.15	1.15 + .21 = 1.36
Theoretical No. of Workers, b = (a)/ Cycle Time	3.15	4	4.1	8.65	9.45	0.8	2.45	5.75	6.8

Task	A	A+B	C	C + D	C + D + E	E	E + F	E + F + G	E + F + G + H
Actual No. of Workers, c	4	4	5	9	10	1	3	6	7
Workers Utilization (%) = (b/c) × 100	78.75%	100%	82%	96.11%	94.5%	80%	81.67%	95.83%	97.14%
	Workstation 1		Workstation 2					Workstation 3	

The next task to be considered is Workstation B. Total task time now for A + B is 0.63 + 0.17 = 0.8 minutes. Theoretical and actual number of workers are four each, and the workers utilization is 100%. As the utilization is 100%, there is no need to add tasks further and Workstation 1 includes the tasks A and B with the requirement of four workers.

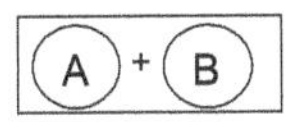

Theoretical number of workers = 4
Actual number of workers = 4

Workstation 1

To begin for Workstation 2, the next task is C with time 0.82 minutes.

$$\text{The theoretical number of workers} = \frac{0.82}{0.2} = 4.1$$

Actual number of workers would be the next higher integer, i.e. five, and the worker utilization is,

$$= \frac{4.1}{5} \times 100 = 82\%$$

The next task to be added is either D or E as per the precedence diagram. Let us say D (in alphabetical order). Following the procedure as discussed before, the theoretical and actual number of workers are 8.65 and nine, respectively. The worker utilization is 96.11%. By adding the task E, utilization is 94.5%. As the worker utilization is decreased by an addition of task E, this task is excluded from the current workstation. Now Workstation 2 has the tasks C and D.

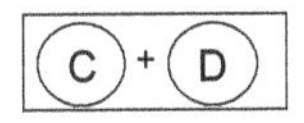

Theoretical number of workers = 8.65
Actual number of workers = 9

Workstation 2

Task E was excluded for Workstation 2 and it becomes the first task for Workstation 3, and the similar procedure is repeated. Now tasks are added from F to H, one by one, and the worker utilization is observed to be increasing continuously to 97.14%. As no other task remains to be considered, Workstation 3 will consist of four tasks—E, F, G and H—with the worker utilization 97.14%.

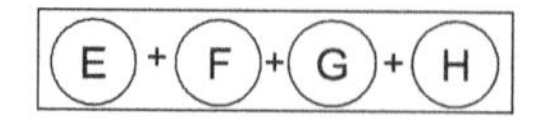

Theoretical number of workers = 6.8
Actual number of workers = 7

Workstation 3

Theoretical number of workers for the entire assembly line

= theoretical number of workers for workstations 1, 2 and 3
= 4 + 8.65 + 6.8
= 19.45, which is the number obtained in part 3. of the current example.

Actual number of workers for the entire assembly line

= actual number of workers for workstations 1, 2 and 3
= 4 + 9 + 7
= 20

5. Efficiency of the proposed solution

$$= \frac{\text{Theoretical total number of workers}}{\text{Actual total number of workers}}$$

$$= \frac{19.45}{20}$$

= 0.9725
= 97.25%

2 PLANNING FOR PROCESS LAYOUT

The main objective in designing the process or functional layout is to minimize the total distance travelled by the material or to minimize material handling costs. Figure 4 provides a brief classification of the methods and computer programmes available for analyzing and developing the process type of layout.

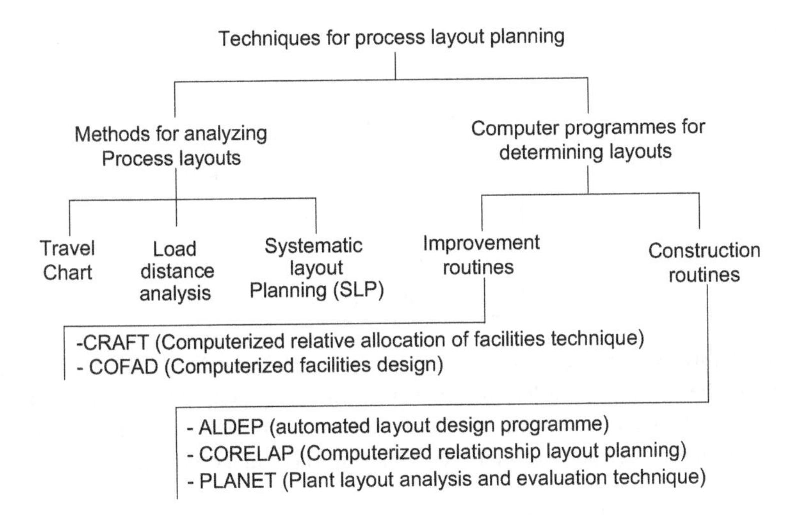

FIGURE 4 Classification of techniques for planning the process layouts.

A	B	C
O	E	D

FIGURE 5 An initial idea of the layout.

2.1 Travel Chart

The travel chart is used to analyze the movement of material/components from department to department. Department refers to a place or location where either similar machines are installed or a similar kind of service is provided. The travel chart method will provide a good solution in the form of the arrangement of departments, which is subject to further analysis and modification. This will take into account an assessment of flow or movement of parts or components in a specified period and the departments between which the movement is greater, and those that are to be positioned together or closer to each other as much as possible.

Example 3

A job shop is in the process of planning for the layout consisting of five departments A, B, C, D and E. An idea is to place the departments in the order as shown in Figure 5, where O is the space reserved for administrative office.

Estimated production quantities per week and the movement order of five products are shown as follows:

Product No.	Production Quantity	Movement Order
1	20	A-C-B-D-E
2	30	D-A-B-C
3	25	A-D
4	60	C-D-E-A
5	40	B-C-D

Product 1 moves from department A to C, then via B and D, it finally moves to E. Similarly, the movement orders of other products are given. Demand of the Product 3 is expected to increase. Assume that the areas of each department are equal. Movement details of each product are entered in the matrix as follows:

20 Nos. of product No.1 travel

1. From department A to C
2. From department C to B
3. From department B to D
4. From department D to E

This is shown in the matrix as follows:

To → / ↓From	A	B	C	D	E
A	—		20		
B		—		20	
C		20	—		
D				—	20
E					—

In this manner, the movement details of the remaining four products are entered in the matrix, and finally, the travel chart is as shown in Figure 6.

To → / ↓From	A	B	C	D	E
A	—	30	20	25	
B		—	30 + 40 = 70	20	
C		20	—	60 + 40 =100	
D	30			—	20 + 60 = 80
E	60				—

FIGURE 6 Travel chart.

From Figure 6, the movements between two departments are as follows:

A-B	30
A-C	20
A-D	25*
B-C	70**
B-D	20
C-B	20**
C-D	100
D-A	30*
D-E	80
E-A	60

Note: * is used to differentiate between 'D-A/A-D' and 'B-C/C-B'

As we are interested in the total flow of material between departments, it is immaterial whether the products move from A to D or from D to A. Therefore, 25* and 30* may be added to get the total flow between D-A/A-D. Similarly between B-C/C-B, the total movement is 70** + 20** = 90. Accordingly, the total movement between departments:

A-B	30
A-C	20
A-D	55
B-C	90
B-D	20
C-D	100
D-E	80
E-A	60

Arranging in the decreasing order of the total movement:

C-D	100
B-C	90
D-E	80
E-A	60
A-D	55
A-B	30
A-C	20
B-D	20

Referring to Figure 5, C-D, B-C and D-E are closer to each other. As the preceding list is in the order of priority for the closeness of departments, an initial idea of the layout is suitable with reference to C-D, B-C and D-E. E-A are also comparatively close to each other. But A-D are far from each other, whereas A-B are closer. A-D must be given the priority over A-B, as the total flow between A-D is 55, which is much higher than 30 between A-B. Further demand of Product 3 is expected to increase, and the path followed by this product is A-D. It is reasonable to interchange the location of B and D, which is shown in Figure 7.

A	D	C
O	E	B

FIGURE 7 Proposed layout.

In this method, the distance between departments is not taken into consideration. Either it may be included in the travel chart analysis or it may be considered in the load-distance method, which is explained in the next section.

2.2 Load-Distance Analysis

In many situations, the objective is to compare two or more alternative layouts. If some alternatives are available, then depending on the movement of various products, the total distance travelled by all the items to be manufactured is computed for the alternative layouts. The preferred layout is obviously the one in which the total distance travelled by the items is the shortest.

Example 4

After careful consideration, an organization has found the two alternative layouts, X and Y, as shown below:

LAYOUT X

A	B	E
D	C	F

LAYOUT Y

F	B	E
A	C	D

Estimated production quantity per week and the movement order of three items to be produced are:

Item No.	Production Quantity	Movement Order
1	200	A-B-D-F-E
2	150	D-E-F-C-A
3	100	F-C-B-A-E

Distances in metres between departments have been provided for both the layouts:

Department (From-To)	Distance (m)	
	Layout X	Layout Y
A-B/B-A	10	20
B-D	20	20
D-F	20	30
F-E/E-F	10	20
D-E	30	10
F-C	10	20
C-A	20	10
C-B	10	10
A-E	20	30

Now the distance moved by each product for both the layout alternatives are computed:

Item	Movement Order	Distance Moved by Each Item (m)	
		Layout X	Layout Y
1	A-B-D-F-E	10 + 20 + 20 + 10 = 60	20 + 20 + 30 + 20= 90
2	D-E-F-C-A	30 + 10 +10 +20 = 70	10 + 20 + 20 + 10 = 60
3	F-C-B-A-E	10 + 10 + 10 + 20 = 50	20 + 10 + 20 + 30 = 80

Distance travelled by all the quantities of each item per week can be obtained as:

Item	Quantity	Distance Moved (m)	
		Layout X	Layout Y
1	200	60 × 200 = 12,000 m	90 × 200 = 18,000 m
2	150	70 × 150 = 10,500 m	60 × 150 = 9,000 m
3	100	50 × 100 = 5,000 m	80 × 100 = 8,000 m

Total movement (m) for each alternative layout:

Layout X: 12,000 + 10,500 + 5,000 = 27,500 m
Layout Y: 18,000 + 9,000 + 8,000 = 35,000 m

As the total movement for layout X is less, it is the preferred layout between alternatives.

2.3 Systematic Layout Planning (SLP)

In certain situations, when the flow of materials between departments is not the only significant factor on which a layout decision can be based, systematic layout planning may be useful. Depending on several factors such as common equipment, common workers, normal flow sequence, better supervision, etc., closeness ratings are developed between departments. Considering the closeness rating, the two departments may need to be placed together because the pair of departments has absolutely necessary rating. But another pair of departments may have an undesirable rating, and these should be located far from each other. Intermediate ratings are also developed based on the various factors. Initial layout is considered, and through trial and error, it is modified by placing departments on the basis of their closeness ratings, subject to the space constraints.

2.4 CRAFT (Computerized Relative Allocation Facilities Technique)

As shown in Figure 4, this is an improvement routine. Improvement routines are those algorithms which require an initial layout of facilities. The objective of the CRAFT is to minimize the total material handling cost. In this heuristic procedure, an improvement in the layout is made by interchanging the departments and comparing the costs. The method is discussed briefly as follows:

1. Input details are: (a) initial layout, (b) cost of material handling per unit distance, and (c) movement data. Compute the total handling cost corresponding to an initial layout.
2. Compute the cost of handling with each pair-wise interchange of the centroids of departments, and select the lowest cost.
3. If the least cost obtained in 2. is less than the cost of previous layout, make the actual interchange of the selected pair of departments and compute the cost of the revised layout.
4. Repeat the procedure from 2. onwards as long as an improvement in the cost is observed.

Following are the assumptions in the CRAFT:

1. Distance between two departments is the rectilinear distance between the centroids of the departments. For example, two departments *A* and *B* are adjacent to each other as shown below:

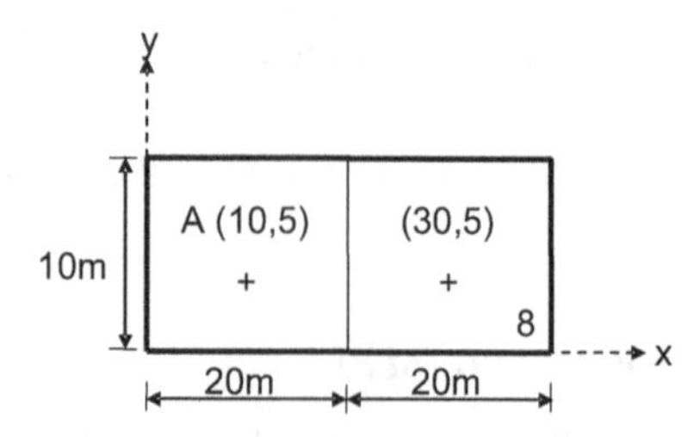

The bottom most horizontal line may be considered as X-axis and the left most vertical line is treated as Y-axis in the diagram.

Coordinates of the centroids of A and B, shown by +, are (10, 5) and (30, 5). The distance between centroids is 20 m, which is obtained as:

$$|10 - 30| + |5 - 5| = |-20| + |0| = 20 + 0 = 20\text{m}$$

(ignoring the negative sign of $|-20|$).

If centroids of the department A and B are denoted as (x_A, y_A), and (x_B, y_B), then the distance between A and B is considered as:

$$d_{AB} = |x_A - x_B| + |y_A - y_B|$$

In case, the departments A and B are as shown below then the centroids are $(x_A, y_A) = (5,5)$ and $(x_B, y_B) = (20, 10)$.

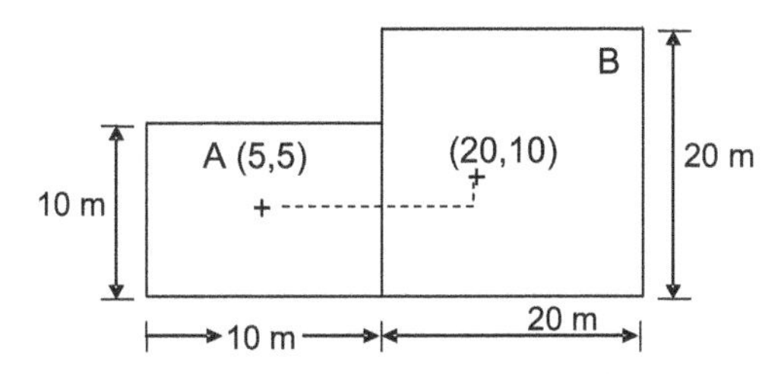

As the distance between two departments is assumed to be the rectilinear distance (shown by dotted lines) between centroids,

$$\begin{aligned} d_{AB} &= |5 - 20| + |5 - 10| \\ &= |-15| + |-5| \\ &= 15 + 5 = 20 \text{ m.} \end{aligned}$$

If a department does not have the rectangular or square shape, then the centroid of the department is obtained as follows:

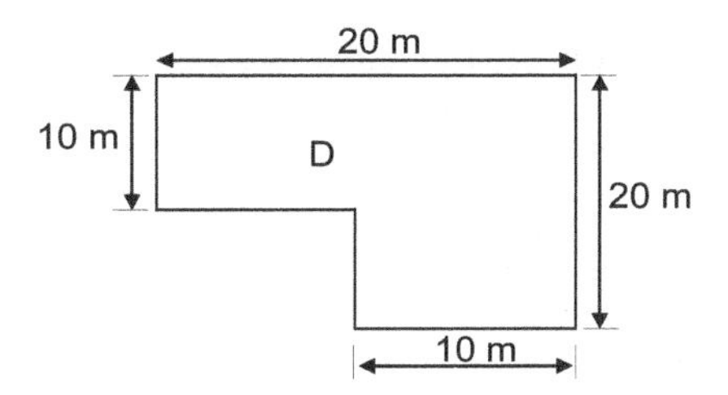

A department D is shown and an objective is to find out its centroid. The whole area should be divided in such a way that the regular shapes are obtained as shown in the following figure:

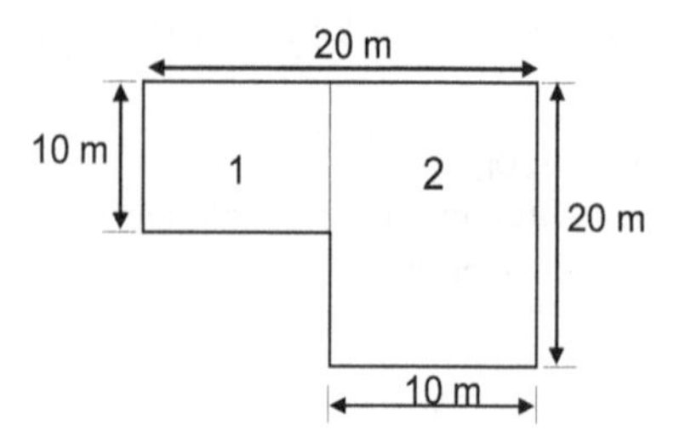

Area of portion 1, $a_1 = 10 \times 10 = 100\ m^2$
Centroid of portion 1, $(x_1, y_1) = (5, 15)$
Similarly for portion 2, $a_2 = 200\ m^2$ and $(x_2, y_2) = (15, 10)$.
The centroids of department D, (x_D, y_D) are given as,

$$x_D = \frac{(a_1 \times x_1) + (a_2 \times x_2}{a_1 + a_2}$$

$$= \frac{(100 \times 5) + (200 \times 15}{100 + 200}$$

$$= \frac{500 + 3000}{300}$$

$$= 11.67$$

and $$y_D = \frac{(a_1 \times y_1) + (a_2 \times y_2}{a_1 + a_2}$$

$$= \frac{(100 \times 15) + (200 \times 10}{100 + 200}$$

$$= 11.67$$

2. Interchange between the departments is based on the following criteria:
   a. Departments have equal area.
   b. Departments are adjacent (i.e. either a border is shared by the departments completely or partially).

Interchange is possible when either criteria a. is satisfied or criteria b. is satisfied, as well as when both a. and b. are fulfilled.

3. While making actual interchanges between the departments, if it is not possible to retain the shape of any of the department, then the shape may change. However, obviously, the area of the department should remain the same.

Example 5

In order to implement the CRAFT, an initial layout is given as shown in Figure 8.

Four departments—A, B, C and D—are to be located in the total available area, $80 \times 40 = 3{,}200$ m^2. Load *Lij* to be transported from one department i to another j in a week is given in the movement matrix as shown in Figure 9.

Suppose that, instead of transportation of one item at a time, a number of items are moving between departments on a single trip; then the movement matrix may also refer to the number of trips for material handling per week or any other specified time period.

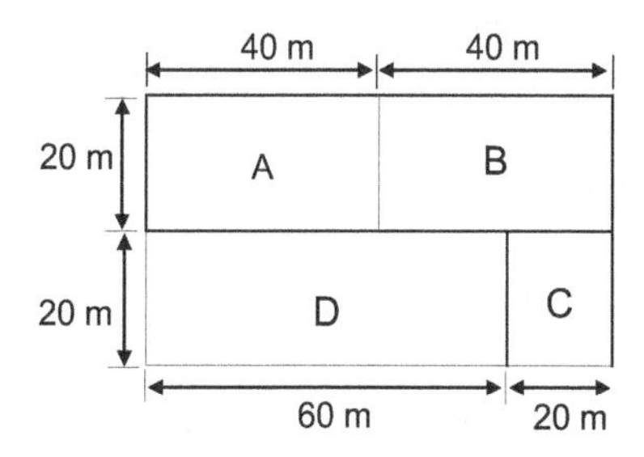

FIGURE 8 Initial layout to implement CRAFT.

From ↓ \ To →	A	B	C	D
A		1	3	2
B	4		2	3
C	2	3		4
D	0	1	2	

FIGURE 9 Movement matrix.

As discussed before, the centroids of departments may be obtained as follows:

x_A, y_A	x_B, y_B	x_C, y_C	x_D, y_D
20, 30	60, 30	70,10	30,10

Distance between the department A and B,

$$d_{AB} = |x_A - x_B| + |y_A - y_B|$$
$$= |20 - 60| + |30 - 30|$$
$$= 40 \text{ m}$$

Similarly, the distances between each pair of departments are computed, and these are shown in Figure 10.

As the distance between A-B, A-C, A-D is similar to B-A, C-A, D-A, respectively, the first row on the right side of the diagonal is similar to the first column on the left side of the diagonal in Figure 10. Similarly, the second and third column on the left side will be the same as second and third row on the right side of diagonal, respectively.

If there are n departments (n = 4 in this example),

$$\text{Total movement (m)} = \sum_{i=1}^{n} \sum_{j=1}^{n} Lij \; dij$$

$$\text{Total cost of transportation} = \sum_{i=1}^{n} \sum_{j=1}^{n} Lij \; dij \; Kij$$

Where K_{ij} is the cost per metre of transportation from the department *i* to the department *j*.

From \ To	A	B	C	D
A		40	70	30
B	40		30	50
C	70	30		40
D	30	50	40	

FIGURE 10 Distance between departments, in metres.

If the cost per unit m is constant, say K, irrespective of the pair of departments,

$$\text{the total material handing/transportation cost} = K \sum_{i=1}^{n} \sum_{j=1}^{n} Lij \;\; dij$$

As K is constant, minimizing the total cost would be similar to minimizing the total movement/distance travelled. For simplicity, K is assumed as Re. 1 per metre so that the numerical value of total cost will also be similar to the total distance moved.

For ease in computation, Figure 9 is superimposed on Figure 10 as shown in Figure 11. The load to be moved or the number of trips per period, Lij, are shown in left comer of each cell.

(Note: If Kij is different for the different pair of departments, then it can be written in the right corner of each cell. Total cost of transportation will be the sum of multiplication of three numbers in each cell. In the present example, as K is constant i.e. K = Re. 1 per metre distance, it is ignored.)

$$\begin{aligned}\text{Total cost of the initial layout} &= 1 \times [(40 \times 1) + (70 \times 3) + (30 \times 2) \\ &\quad + (40 \times 4) + (30 \times 2) + (50 \times 3) \\ &\quad + (70 \times 2) + (30 \times 3) + (40 \times 4) \\ &\quad + (30 \times 0) + (50 \times 1) + (40 \times 2)] \\ &= \text{Rs. } 1{,}200.00\end{aligned}$$

Now the procedure is to interchange the centroids of departments and compare the least cost corresponding to an interchange with Rs. 1,200. If it is less than Rs. 1,200, then make the actual interchange and compute the cost.

→ To From ↓	A	B	C	D
A		1 40	3 70	2 30
B	4 40		2 30	3 50
C	2 70	3 30		4 40
C	0 30	1 50	2 40	

FIGURE 11 Distance and movement between the departments.

Cycle 1

Interchange between departments may be as follows:

A-B
A-C
A-D
B-C
B-D
C-D

As the interchange between A and C does not satisfy any of the criteria, i.e., A and C do not have equal area and they are also not adjacent, therefore A-C will be dropped. A and B are adjacent, as well, as both have equal area i.e. 800 m^2 and A-B will be considered for the interchange. Remaining departments will also be interchanged because these are adjacent. While interchanging the departments, initially only centroids will be interchanged. An actual interchange will be made only for the pair which has the potential of reduction in cost.

For interchange of the centroids, the reference set of centroids for existing (initial) layout is as shown in Figure 12.

Interchange A-B As discussed, the centroid of A and B will take the place of each other. Now the position of centroids for this interchange is as follows:

x_A, y_A	x_B, y_B	x_C, y_C	x_D, y_D
60, 30	20, 30	70,10	30,10

Distance between each pair of the departments is obtained and the movement data are superimposed as discussed before:

x_A, y_A	x_B, y_B	x_C, y_C	x_D, y_D
20, 30	*60, 30*	*70,10*	*30,10*

FIGURE 12 Reference set of centroids for Cycle 1.

From \ To	A	B	C	D
A		1 40	3 30	2 50
B	4 40		2 70	3 30
C	2 30	3 70		4 40
C	0 50	1 30	2 40	

$$\begin{aligned}\text{The Corresponding Cost} &= (40 \times 1) + (30 \times 3) + (50 \times 2)\\ &\quad + (40 \times 4) + (70 \times 2) + (30 \times 3)\\ &\quad + (30 \times 2) + (70 \times 3) + (40 \times 4)\\ &\quad + (50 \times 0) + (30 \times 1) + (40 \times 2)\\ &= \text{Rs. } 1{,}160.00\end{aligned}$$

Interchange A-D From the reference set of centroids shown in Figure 12, interchange the centroid of A and D as follows:

x_A, y_A	x_B, y_B	x_C, y_C	x_D, y_D
30, 10	60, 30	70,10	20,30

From \ To	A	B	C	D
A		1 50	3 40	2 30
B	4 50		2 30	3 40
C	2 40	3 30		4 70
C	0 30	1 40	2 70	

And the corresponding cost = Rs. 1,240

Interchange B-C Again refer to Figure 12 and the interchange is as follows:

x_A, y_A	x_B, y_B	x_C, y_C	x_D, y_D
20, 30	70, 10	60,30	30,10

From \ To	A	B	C	D
A		1 / 70	3 / 40	2 / 30
B	4 / 70		2 / 30	3 / 40
C	2 / 40	3 / 30		4 / 50
C	0 / 30	1 / 40	2 / 50	

Cost = Rs. 1,220

Interchange B-D

x_A, y_A	x_B, y_B	x_C, y_C	x_D, y_D
20, 30	30, 10	70,10	60,30

From \ To	A	B	C	D
A		1 / 30	3 / 70	2 / 40
B	4 / 30		2 / 40	3 / 50
C	2 / 70	3 / 40		4 / 30
C	0 / 40	1 / 50	2 / 30	

Cost = Rs. 1,160

Interchange C-D

x_A, y_A	x_B, y_B	x_C, y_C	x_D, y_D
20, 30	60, 30	30,10	70, 10

To From	A	B	C	D
A		1 40	3 30	2 70
B	4 40		2 50	3 30
C	2 30	3 50		4 40
D	0 70	1 30	2 40	

Cost = Rs. 1,100

Summary of the cost corresponding to interchange is given as follows:

Interchange:	**(A-B)**	**(A-D)**	**(B-C)**	**(B-D)**	**(C-D)**
Cost (Rs):	1,160	1,240	1,220	1,160	1,100

Minimum cost of Rs. 1,100 corresponds to the change of centroids of departments C-D. As this is less than the cost of the existing layout i.e. Rs. 1,200, make the actual interchange of department C-D. Refer to Figure 8 and interchange the location of C and D as shown in Figure 13.

While making actual interchange, centroids of the departments may change, therefore recalculate the centroids considering Figure 13 and then obtain the actual total cost.

x_A, y_A	x_B, y_B	x_C, y_C	x_D, y_D
20, 30	60, 30	10,10	50, 10

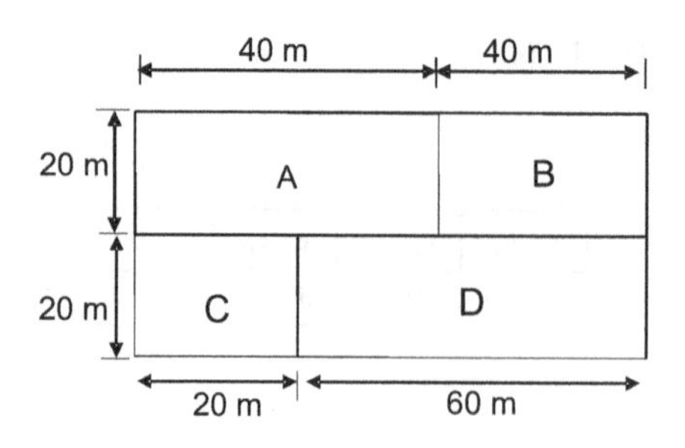

FIGURE 13 Revised layout at the end of Cycle 1.

From \ To	A	B	C	D
A		1 / 40	3 / 30	2 / 50
B	4 / 40		2 / 70	3 / 30
C	2 / 30	3 / 70		4 / 40
D	0 / 50	1 / 30	2 / 40	

Total actual cost of the revised layout = Rs. **1,160**

If this cost is less than the cost of existing layout, then only retain Figure 13 as the revised layout. As in this case, the total actual cost Rs. 1,160 is less than Rs. 1,200, the revised layout which will be considered for further computation is retained, and it will replace the initial layout. Now call it the existing layout. For comparing the cost and interchange between the departments, the recently obtained layout, centroids and cost will act as reference. Now the next cycle is to be followed.

Cycle 2

Out of the interchanges A-B, A-C, A-D, B-C, B-D and C-D, since the departments B and C neither have equal area nor are adjacent, B-C is not to be considered. As the interchange C-D is recently made, if we try to again interchange C-D, the initial layout will be obtained, which will yield no improvement in the cost. Therefore, the recently interchanged department pair is not to be considered.

For Cycle 2, four interchanges are to be explored, viz. A-B, A-C, A-D and B-D. Reference set of centroids for the existing layout is as follows:

x_A, y_A	x_B, y_B	x_C, y_C	x_D, y_D
20, 30	60, 30	10,10	50, 10

Interchange A-B

x_A, y_A	x_B, y_B	x_C, y_C	x_D, y_D
60, 30	20, 30	10,10	50, 10

To / From	A	B	C	D
A		1 40	3 70	2 30
B	4 40		2 30	3 50
C	2 70	3 30		4 40
C	0 30	1 50	2 40	

Cost = Rs. 1,200

Interchange A-C

x_A, y_A	x_B, y_B	x_C, y_C	x_D, y_D
10, 10	60, 30	20,30	50, 10

To / From	A	B	C	D
A		1 70	3 30	2 40
B	4 70		2 40	3 30
C	2 30	3 40		4 50
C	0 40	1 30	2 50	

Cost = Rs. 1,200

Interchange A-D

x_A, y_A	x_B, y_B	x_C, y_C	x_D, y_D
50, 10	60, 30	10,10	20, 30

From \ To	A	B	C	D
A		1 / 30	3 / 40	2 / 50
B	4 / 30		2 / 70	3 / 40
C	2 / 40	3 / 70		4 / 30
	0	1	2	

Cost = Rs. 1,140

Interchange B-D

x_A, y_A	x_B, y_B	x_C, y_C	x_D, y_D
20, 30	50, 10	10,10	60, 30

From \ To	A	B	C	D
A		1 / 50	3 / 30	2 / 40
B	4 / 50		2 / 40	3 / 30
C	2 / 30	3 / 40		4 / 70
D	0 / 40	1 / 30	2 / 70	

Cost = Rs. 1,220

Out of these interchanges, the minimum cost of Rs. 1,140 corresponds to A-D, which is less than the cost of the existing layout (Rs. 1,160). Actual interchange between

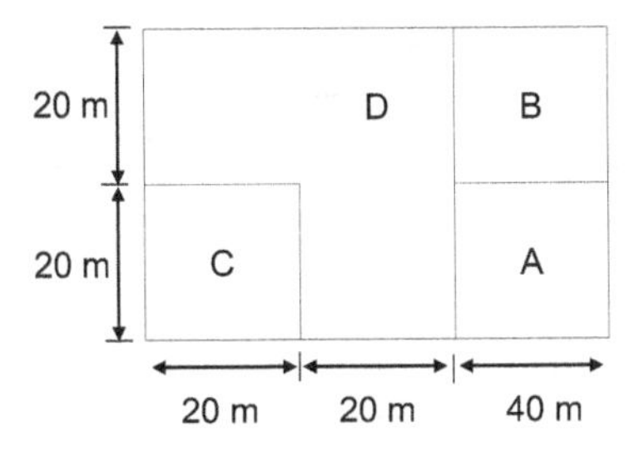

FIGURE 14 Revised layout at the end of Cycle 2.

A and D in the existing layout (Figure 13) will yield the revised layout at the end of Cycle 2, as shown in Figure 14.

Referring to Figure 13, A has taken the place of D, as shown in Figure 14. In the remaining unoccupied portion, D is accommodated. The area of D is same as before, but the shape is changed as shown following:

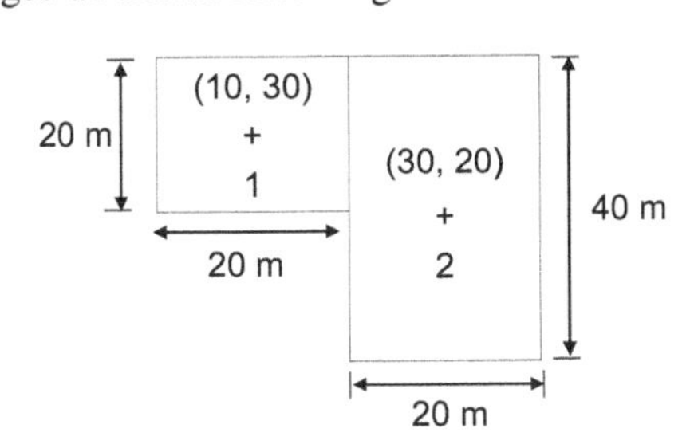

The whole area of department may be divided into two regular shapes, 1 and 2, in order to obtain the centroid (x_D, y_D) as explained before. The centroids of portions 1 and 2 are shown by + symbols and the values are (10, 30) and (30, 20), respectively.

$$x_D = \frac{a_1 \ x_1 \ + a_2 \ x_2}{a_1 \ + a_2}$$

Where $a_1 = 400 \text{ m}^2, a_2 = 800\text{m}^2, x_1 = 10, x_2 = 30$

and $x_D = \dfrac{(400 \times 10) + (800 \times 30)}{1200} = 23.3$

$$y_D = \frac{a_1 \, y_1 \ + a_2 \, y_2}{a_1 \ + \ a_2}$$

Where $y_1 = 30, y_2 = 20$

and $y_D = \dfrac{(400 \times 30) + (800 \times 20)}{1200} = 23.3$

Now with reference to the Figure 14, the centroids are as follows:

x_A, y_A	x_B, y_B	x_C, y_C	x_D, y_D
60, 10	60, 30	10,10	23.33, 23.33

From \ To	A	B	C	D
A		1 20	3 50	2 50
B	4 20		2 70	3 43.44
C	2 50	3 70		4 26.66
D	0 50	1 43.34	2 26.66	

Total actual cost of the revised layout = Rs. 1,133.32

As it is less than the cost of the existing layout (Rs. 1,160), Figure 14 is retained as the revised layout.

Cycle 3

As A-D is recently interchanged, it will not be considered.

A-C and B-C do not satisfy either of the criteria for interchange. Possibilities to be explored confine to the three pair-wise interchanges A-B, B-D and C-D.

Interchange A-B Refer to the recently obtained centroids for the revised layout and interchange them for A-B.

x_A, y_A	x_B, y_B	x_C, y_C	x_D, y_D
60, 30	60, 10	10,10	(23.33, 23.33)

From \ To	A	B	C	D
A		1 20	3 70	2 43.34
B	4 20		2 50	3 50
C	2 70	3 50		4 26.66
D	0 43.34	1 50	2 26.66	

Cost = Rs. 1,146.64

Interchange B-D

x_A, y_A	x_B, y_B	x_C, y_C	x_D, y_D
60, 10	23.33, 23.33	10,10	60, 30

From \ To	A	B	C	D
A		1 50	3 50	2 20
B	4 50		2 26.66	3 43.34
C	2 50	3 26.66		4 70
D	0 20	1 43.34	2 70	

Cost = Rs. 1,266.66

Interchange C-D

x_A, y_A	x_B, y_B	x_C, y_C	x_D, y_D
60, 10	60, 30	(23.33, 23.33)	(10, 10)

Cost = Rs. 1,106.66

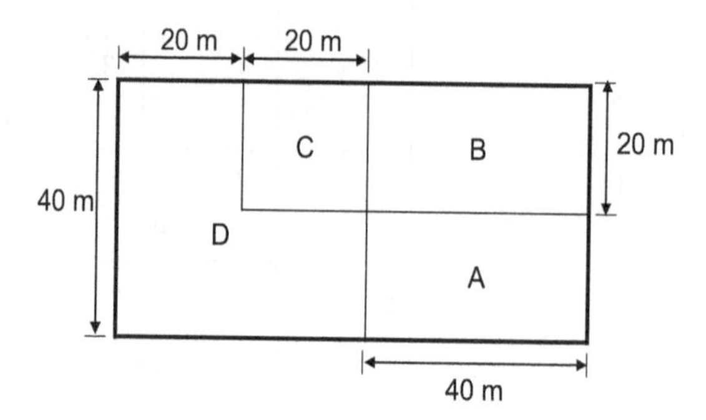

FIGURE 15 Revised layout at the end of Cycle 3

Minimum cost obtained is Rs. 1,106.66 corresponding to the interchange C-D which is less than the cost of existing layout (Rs. 1,133.32), therefore make the actual interchange C-D, as shown in Figure 15.

As discussed in Cycle 2, the centroids are obtained as:

x_A, y_A	x_B, y_B	x_C, y_C	x_D, y_D
60, 10	60, 30	30, 30	16.66, 16.66

From \ To	A	B	C	D
A		1 20	3 50	2 50
B	4 20		2 30	3 56.68
C	2 50	3 30		4 26.68
D	0 50	1 56.68	2 26.68	

Total actual cost of the revised layout = Rs. 986.80

As it is less than the cost of existing layout (Rs. 1,133.32), Figure 15 is retained as the revised layout.

Cycle 4

Since C-D is recently interchanged and A-C and B-D do not satisfy either of the criteria, these pairs will not be considered.

Interchange A-B

x_A, y_A	x_B, y_B	x_C, y_C	x_D, y_D
60, 30	60, 10	30, 30	16.66, 16.66

From \ To	A	B	C	D
A		1 20	3 30	2 56.68
B	4 20		2 50	3 50
C	2 30	3 50		4 26.68
D	0 56.68	1 50	2 26.68	

Cost = Rs. 973.44

Interchange A-D

x_A, y_A	x_B, y_B	x_C, y_C	x_D, y_D
16.66, 16.66	60, 30	30, 30	60, 10

From \ To	A	B	C	D
A		1 56.68	3 26.68	2 50
B	4 56.68		2 30	3 20
C	2 26.68	3 30		4 50
D	0 50	1 20	2 50	

Cost = Rs. 1,046.80

Interchange B-C

x_A, y_A	x_B, y_B	x_C, y_C	x_D, y_D
60, 10	30, 30	60, 30	16.66

From \ To	A	B	C	D
A		1, 50	3, 20	2, 50
B	4, 50		2, 30	3, 26.68
C	2, 20	3, 30		4, 56.68
D	0, 50	1, 26.68	2, 56.68	

Cost = Rs. 1,046.80

The least cost is Rs. 973.44 corresponding to A-B, and it is also less than Rs. 986.80 (the cost of the existing layout); therefore, make actual interchange as shown in Figure 16.

With the actual interchange of departments, the centroids are similar to the situation when only centroids of A and B were interchanged. Therefore, the total actual cost of the revised layout = Rs. 973.44.

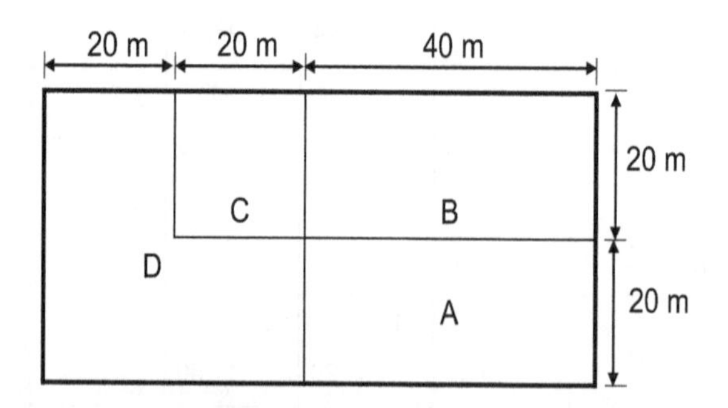

FIGURE 16 Revised layout at the end of Cycle 4.

Cycle 5

As A-B is recently interchanged and A-D and B-C need not be considered, possibilities for further improvement in the cost are to be explored for interchange A-C, B-D and C-D as follows:

Interchange A-C

x_A, y_A	x_B, y_B	x_C, y_C	x_D, y_D
30, 30	60, 10	60, 30	16.66, 16.66

From \ To	A	B	C	D
A		1 50	3 30	2 26.68
B	4 50		2 20	3 50
C	2 30	3 20		4 56.68
D	0 26.68	1 50	2 56.68	

Cost = Rs. 1,093.44

Interchange B-D

x_A, y_A	x_B, y_B	x_C, y_C	x_D, y_D
60, 30	16.66, 16.66	30, 30	60, 10

From \ To	A	B	C	D
A		1 56.68	3 30	2 20
B	4 56.68		2 26.68	3 50
C	2 30	3 26.68		4 50
D	0 20	1 50	2 50	

Cost = Rs. 1,106.80

Interchange C-D

x_A, y_A	x_B, y_B	x_C, y_C	x_D, y_D
60, 30	60, 10	(16.66, 16.66)	30.30

From \ To	A	B	C	D
A		1 / 20	3 / 56.68	2 / 30
B	4 / 20		2 / 50	3 / 50
C	2 / 56.68	3 / 50		4 / 26.68
D	0	1	2	

Cost = Rs. 1,053.48

Minimum cost is Rs. 1,053.48, which is more than the cost of the existing layout (Rs. 973.44). Hence, the process of further computation is stopped and Figure 16 represents the proposed layout with the total cost of Rs. 973.44.

2.5 COFAD (Computerized Facilities Design)

Computerized facilities design (COFAD) is further extension of CRAFT. Overall cost also depends on the selection of material handling equipment, in addition to the distance travelled. COFAD considers various options for the material handling equipment from the point of view of equipment utilization and cost. After selecting the equipment, the layout is improved following the procedure of CRAFT.

2.6 ALDEP (Automated Layout Design Programme)

As shown in Figure 4, ALDEP is a construction routine. Unlike improvement routines, an initial layout is not required. A relationship chart is developed which will show the relationship between departments in the form of closeness ratings. Closeness ratings are as follows:

Absolutely Necessary (A) The letter code for 'absolutely necessary' closeness rating is 'A'. For example, if a similar facility or service centre is being used by the processes to be carried out in two departments, then these need to be adjacent to each other.

Especially Important (E)	For example, if the same human resources are needed for two departments, then from the point of view of supervision, these may be considered under closeness rating 'E'.
Important (I)	It may be important to arrange the departments in the same sequence as the normal flow of material takes place.
Ordinary Importance (O)	If it is not necessary to keep two departments together, but the management considers it convenient to place them as near to each other as possible.
Unimportant (U)	If we are indifferent about the location of two or more departments, then it is not a matter of consideration whether two departments should be close to each other or far from each other.
Undesirable (X)	From safety considerations, etc., it may be undesirable to place the two departments adjacent to each other. This rating is denoted by 'X'.

Numerical values attached to the closeness ratings are as follows:

Letter Code	Numerical Value
A	$4^3 = 64$
E	$4^2 = 16$
I	$4^1 = 4$
O	$4^0 = 1$
U	0
X	$-4^5 = -1{,}024$

The first department is selected at random to begin with and then depending on the closeness rating, the next department is chosen to be adjacent to the first department. Suitable sweep width in number of columns needs to be specified.

For example, if a department '*a*' has an area of 10 m^2, then *a* may be shown either as 1× 10 m or 2 × 5 m:

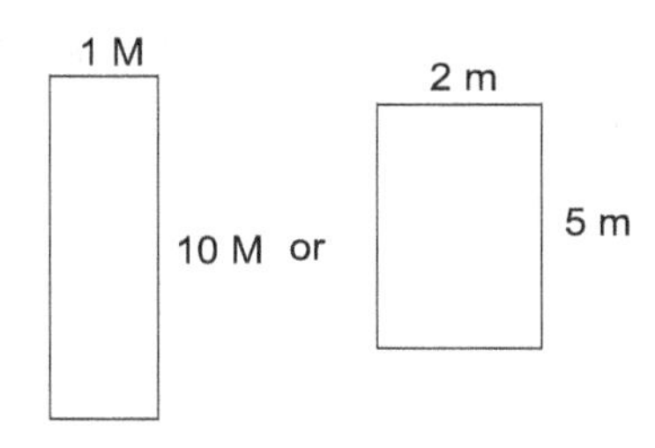

If the sweep width is one column, then starting from the top left comer of the layout, 'a' is printed ten times vertically, as follows (each 'a' represents 1 m^2):

a
a

a
a
a
a
a
a
a
a

If the sweep width is two columns, then it is printed as:

aa
aa
aa
aa
aa

Usually, odd shapes of the departments are obtained in ALDEP, which needs to be modified further.

The basic concept with reference to ALDEP is explained with the help of Example 6.

Example 6

A relationship chart showing the closeness ratings between any two departments is shown following:

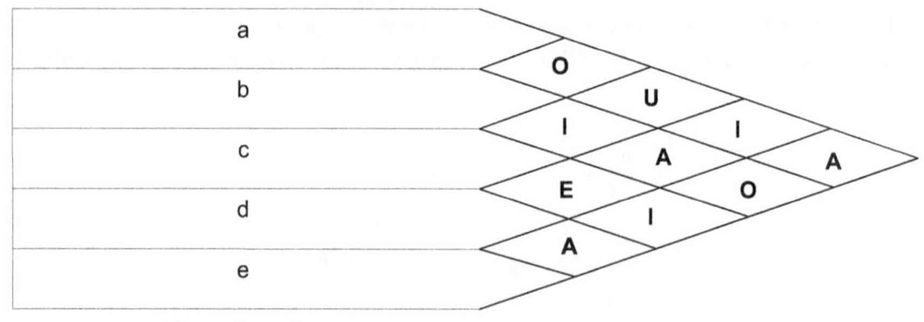

Departments

Five departments—*a*, *b*, *c*, *d* and *e*—are to be located in a layout.

From the relationship diagram, closeness rating between *a* and *b* is O, between *a* and *c* is U, and so on. Obviously, rating between *b* and *a* will also be O, and between c and *a*, it will be U. Closeness rating for the pair of departments are summarized as follows:

Pair of Departments	Closeness Rating
a-b/b-a	O
a-c/c-a	U
a-d/d-a	I
a-e/e-a	A
b-c/c-b	I
b-d/d-b	A
b-e/e-b	O
c-d/d-c	E
c -e/e-c	I
d-e/e-d	A

Area of the departments are:

Department:	*a*	*b*	*c*	*d*	*e*
Area in m² (× 100):	9	6	15	18	12

Ignoring (×100) m² which is common to all departments, the total area of the departments/layout is 60 m². Let us say that the total area available is 10 × 6 m², as shown following:

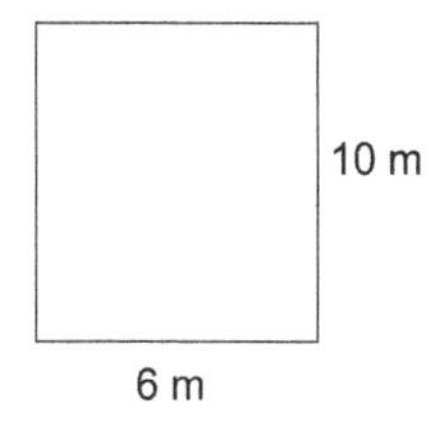

Assume that the sweep width = three columns.

Cycle 1

The first department is selected at random, say *b*. Now the three columns simultaneously will be filled by *b* starting from the top left comer of the layout as represented following:

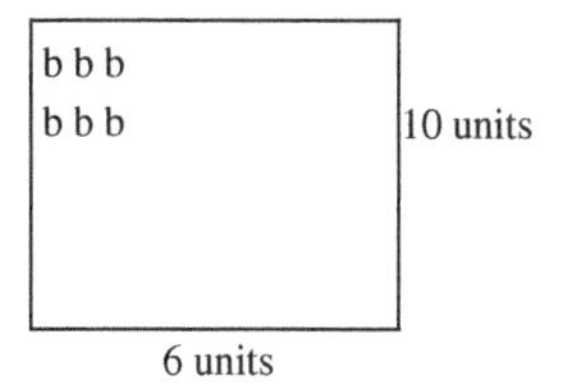

Area of *b* is 6 square units and each b represents 1 square unit. Therefore, the total six *'bs'* in the three columns are accommodated because the sweep width is three columns.

Recently entered department is *b*. The closeness ratings of the remaining departments with b are as follows:

Remaining Departments:	*a*	*c*	*d*	*e*
Rating with *b* :	O	I	A	O

The highest rating is A, corresponding to the department *d*; therefore, *d* is selected next to enter in the layout. (Note: If minimum department preference value is specified as, say, I in terms of the closeness rating and if all ratings are lower than I, then any department would be selected at random.)

Area of *d* is 18 square units, and therefore the six rows would be filled in three column width as represented following:

b	b	b
b	b	b
d	d	d
d	d	d
d	d	d
d	d	d
d	d	d
d	d	d

10 Units

6 units

(out of ten units, in the first three column width, eight units are filled. Two rows are empty in this column width.)

Recently selected department is *d*.

Remaining Departments:	*a*	*c*	*e*
Rating with *d* :	I	E	A

The highest rating is *A*, corresponding to the department *e*, and *e* is selected to enter. Area of *e* is 12 square units and it will require four rows of three column width. After filling two rows in the first sweep width, the bottom of the layout will be reached. Now start from the bottom of the second sweep width of three columns as shown following:

b	b	b
b	b	b
d	d	d

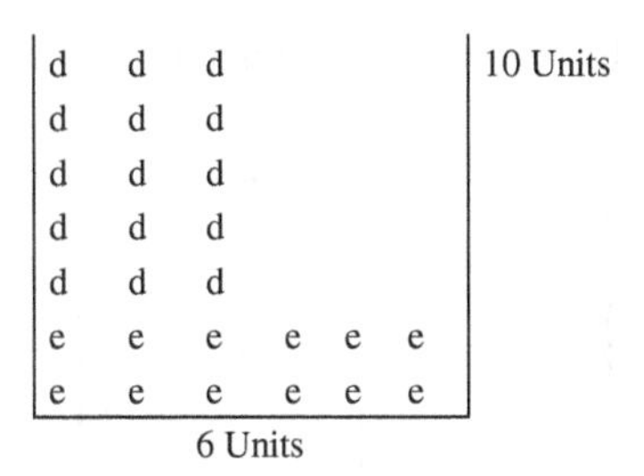

d	d	d			
d	d	d			
d	d	d			
d	d	d			
d	d	d			
e	e	e	e	e	e
e	e	e	e	e	e

10 Units

6 Units

Recently entered department is *e*.

Remaining Departments:	***a***	***c***
Rating with e:	A	I

Department *a* is selected and filled in the layout as shown following:

b	b	b			
b	b	b			
d	d	d			
d	d	d			
d	d	d			
d	d	d	a	a	a
d	d	d	a	a	a
d	d	d	a	a	a
e	e	e	e	e	e
e	e	e	e	e	e

10 Units

6 units

Remaining department is only *c*, which is entered as shown in Figure 17.

b	b	b	c	c	c
b	b	b	c	c	c
d	d	d	c	c	c
d	d	d	c	c	c
d	d	d	c	c	c
d	d	d	a	a	a
d	d	d	a	a	a
d	d	d	a	a	a
e	e	e	e	e	e
e	e	e	e	e	e

10 Units

6 units

FIGURE 17 Proposed layout at the end of Cycle 1.

This proposed layout by ALDEP may be imagined in the following form:

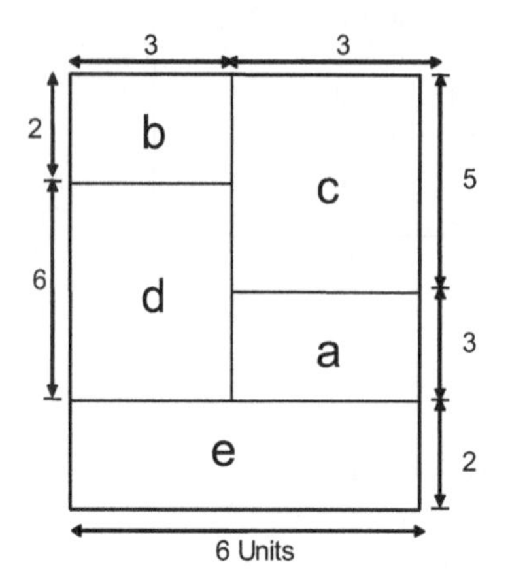

Pairs of the adjacent departments are b-c, b-d, d-c, d-a, d-e, e-a, and a-c. The objective is to maximize the total closeness rating of the pairs of adjacent departments. Total closeness rating for the proposed layout at the end of cycle is obtained as follows:

Pair of Adjacent Departments	Closeness Rating	Value
b-c	I	4
b-d	A	64
d-c	E	16
d-a	I	4
d-e	A	64
e-a	A	64
a-c	U	0
Total closeness rating =		**216**

The process may be repeated many times by considering any other department to enter first in the layout. Total rating is evaluated and the layout with maximum total rating is retained as the solution.

For example, in Cycle 2, *c* may enter first.

Cycle 2

Department *c* is the first in the sequence. Remaining departments are *a*, *b*, *d* and *e*, and their ratings with *c* are U, I, E and I, respectively. Maximum rating is E with respect to the department *d*, and therefore, *d* is selected.

Now the selected department is *d*.

Remaining Departments:	***a***	***b***	***e***
Rating with d:	I	A	A

Both *b* and *e* have maximum rating. Let *e* be selected.

Remaining Departments:	***a***	***b***
Rating with e:	A	O

Department *a* is selected, and then finally *b* will enter. The sequence obtained is c-d-e-a-b. After filling the departments in this sequence, the layout will look like as shown following:

c	c	c	b	b	b
c	c	c	b	b	b
c	c	c	a	a	a
c	c	c	a	a	a
c	c	c	a	a	a
d	d	d	e	e	e
d	d	d	e	e	e
d	d	d	e	e	e
d	d	d	e	e	e
d	d	d	d	d	d

10 units

6 units

This layout may be imagined as follows, where *d* has an odd shape.

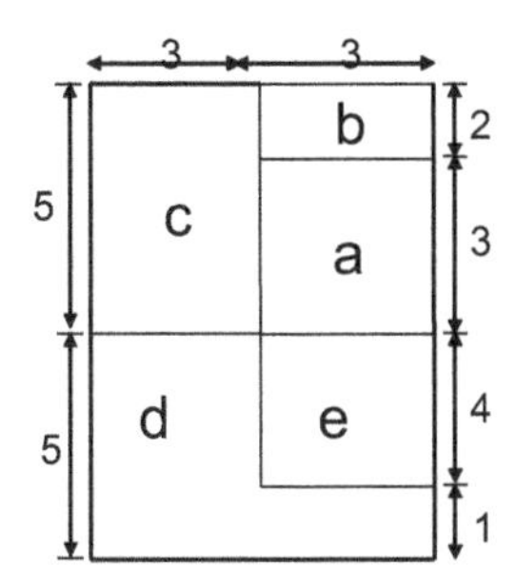

Pair of Adjacent Departments	**Closeness Rating**	**Value**
c-b	I	4
c-a	U	0
c-d	E	16
d-e	A	64
e-a	A	64
a-b	O	1
Total closeness rating =		**149**

The total number of cycles to be computed may be specified. If only two cycles are to be performed, then the maximum value of total rating is 216 and Figure 17 is retained as the proposed solution.

2.7 CORELAP (Computerized Relationship Layout Planning)

CORELAP is also a construction routine. The following closeness rating values are used:

Closeness rating	Value
A	6
E	5
I	4
O	3
U	2
X	1

The first department is not selected at random, but the selection is based on the total closeness rating. Consider the relationship chart of Example 6.

Total closeness rating of department *a*:

= sum of closeness rating of *a* with the remaining departments.
= sum of closeness rating of (a – b), (a – c), (a – d), (a – e)
= O + U + I + A
= 3 + 2 + 4 + 6
= 15

Similarly, the total closeness rating of *b*:

= Sum of closeness rating of (b – a), (b – c), (b – d), (b – e)
= O + I + A + O
= 3 + 4 + 6 + 3
= 16

If there are the total n departments, then for each department, the total closeness rating will be the sum of (n – 1) number of closeness ratings. As there are five departments, for each department, the sum of four closeness ratings will yield the total closeness rating.

The department with the maximum total closeness rating is selected and placed at the centre of the layout. In case of a tie, the department with maximum area is selected first.

3 FORMATION OF CELLS IN THE CELLULAR MANUFACTURING LAYOUT

Cellular manufacturing layout is suitable for the group technology approach. Usually for an existing job shop, some items are to be identified which can be produced in medium quantities. The items are grouped based on their design requirements or operational requirements. Depending on the requirement of the group of items, necessary equipment is installed together in order to form a manufacturing cell.

3.1 Cell Formation

Consider that a job shop has identified six items for which there is keen interest in knowing whether the manufacturing cells can be formed. A cell will comprise certain machines, and each cell will act as a product layout for specified items because the machines are arranged according to the operations to be carried out. Specified items will be completely manufactured in the cell itself.

As an example, the machine-item details are shown in the following matrix:

		Item					
		1	2	3	4	5	6
Machine	1	1	0	0	1	0	1
	2	0	1	1	0	1	0
	3	1	0	1	1	0	1
	4	0	1	1	0	1	0
	5	1	0	0	1	0	1

Five types of machines are there which will be used as per the operations to be performed on items. Third row corresponding to machine 3 is 1 0 1 1 0 1; that means that Machine 3 will be required to produce items 1, 3, 4 and 6. In other words, the items 1, 3, 4, and 6 will need Machine 3 for their processing. Each '1' refers to the requirement of machines for items. Each '0' refers to the non-requirement. Each '0' appears for items 2 and 5 corresponding to Machine 3. Machine 3 is not required to process items 2 and 5, or items 2 and 5 do not need the Machine 3 for manufacturing them.

3.1.1 Column Adjustment

Observe the columns. The items 1, 4 and 6 require similar machines (viz. 1, 3, and 5). Similarly, items 2 and 5 require similar machines. Rearrange these columns to make them adjacent as shown below:

		Item					
		1	4	6	2	5	3
Machine	1	1	1	1	0	0	0
	2	0	0	0	1	1	1
	3	1	1	1	0	0	1
	4	0	0	0	1	1	1
	5	1	1	1	0	0	0

3.1.2 Row Adjustment

Observe the rows. Machines 1, 3, and 5 are required to produce the items (1, 4, and 6) completely. Make them adjacent. Similarly, the machines 2 and 4 are required to produce items 2 and 5 completely. Rearrange the rows as follows:

Item

Machine	1	4	6	2	5	3
1	1	1	1	0	0	0
3	1	1	1	0	0	1
5	1	1	1	0	0	0
2	0	0	0	1	1	1
4	0	0	0	1	1	1

Now two cells can be formed. In one cell, the machines 1, 3, and 5; and in another cell, the machines 2 and 4 can be installed as shown following:

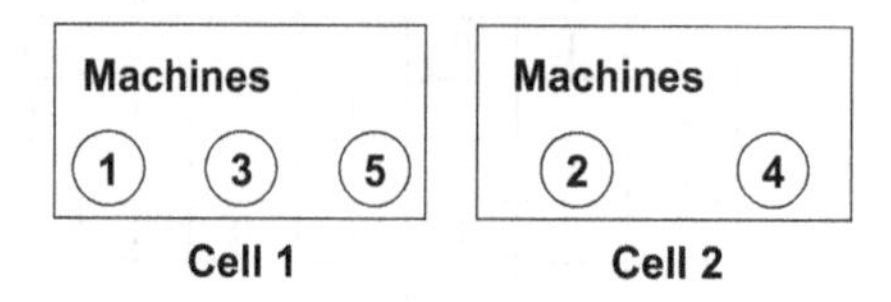

Cell 1 is devoted to the manufacture of items 1, 4 and 6
Cell 2 is used to manufacture the items 2 and 5

Item 3 cannot be grouped with any of the cells because it requires the processing in both the cells. Machine 3 of Cell 1, as well as machines 2 and 4 of Cell 2, are required to manufacture Item 3. Processing of Item 3 may be treated as a job shop or functional layout. Material handling is more for Item 3.

3.2 Bond Energy Algorithm

Cells are also formed with the use of bond energy algorithm. For example, if the machines and items details are as follows:

Item

Machine	1	2	3	4
1	1	0	0	1
2	0	1	1	0
3	1	0	1	1
4	0	1	1	0

Then the bond energy between any two rows, say 1–3, will be the sum of multiplication of the corresponding numbers of row 1 and row 3.

Row 1:	1	0	0	1
Row 3:	1	0	1	1

Bond energy between row 1 and 3:

$= (1 \times 1) + (0 \times 0) + (0 \times 1) + (1 \times 1)$
$= 1 + 0 + 0 + 1$
$= 2$

Similarly, bond energy between row 1 and 4:

$= (1 \times 0) + (0 \times 1) + (0 \times 1) + (1 \times 0)$
$= 0$

Following the same procedure, the bond energy between any two columns, say 1 and 3, will be the sum of multiplication of corresponding number in Column 1 and Column 3.

Column 1	Column 3
1	0
0	1
1	1
0	1

Bond energy between column 1 and column 3:

$= (1 \times 0) + (0 \times 1) + (1 \times 1) + (0 \times 1)$
$= 0 + 0 + 1 + 0$
$= 1$

In this way, the bond energy may be computed for any two rows or for any two columns.

To apply the bond energy algorithm for the formation of cells, first of all, a column is selected at random. Bond energy of this column with each remaining column is computed, and the column with maximum bond energy is chosen to be adjacent to the earlier randomly selected column. A separate matrix is formed by selecting one column after another, following the procedure with the remaining columns. This recently obtained matrix is now considered for the row-wise adjustment, and finally, an attempt is made to form the suitable cells.

Example 7

Consider the following machine-item details:

		Item			
		1	2	3	4
Machine	1	0	1	1	0
	2	0	0	0	1
	3	1	0	0	1
	4	0	1	0	0
	5	1	0	0	1
	6	0	1	1	0

Now, Column 2 is selected at random. In a separate matrix, Column 2 will be arranged as the first column.

Remaining Columns:	1	3	4
Bond Energy of Column:			
2 with the Remaining Columns			
= Bond Energy between the Columns:	2–1	2–3	2–4
Bond Energy:	0	2	0

Maximum bond energy is 2, corresponding to Column 3; therefore, Column 3 is selected next to enter in the matrix.

As the selected column is 3,

Remaining Columns:	1	4
Bond Energy between the Columns:	3–1	3–4
Bond Energy:	0	0

As both columns have equal bond energy with the column 3, select any column; say, 1.

The only remaining column is Column 4; it will be last in the sequence. Sequence of columns in the separately formed matrix is 2-3-1-4 as follows:

		Item			
		2	3	1	4
Machine	1	1	1	0	0
	2	0	0	0	1
	3	0	0	1	1
	4	1	0	0	0
	5	0	0	1	1
	6	1	1	0	0

The preceding matrix will be treated as the reference matrix for a row-wise adjustment.

Select any row, say 4 randomly. The bond energy of row 4 with the remaining rows will be that of pair of rows 4–1, 4–2, 4–3, 4–5 and 4–6.

Remaining Rows:	1	2	3	5	6
Bond Energy with 4:	1	0	0	0	1

Both rows 1 and 6 have the maximum bond energy i.e. 1 with row 4. Select any, say row 1.

Remaining Rows:	2	3	5	6
Bond Energy with Row 1:	0	0	0	2

Maximum bond energy is 2, corresponding to the row 6; therefore the row 6 is selected next.

Remaining Rows:	2	3	5
Bond Energy with the Row 6:	0	0	0

Select any row, say 2.

Remaining Rows:	3	5
Bond Energy with the Row 2:	1	1

Select any row, say 3. The remaining row is row 5 only. The rows selected are in the sequence 4-1-6-2-3-5. The new matrix will be formed considering the adjustment of rows in this sequence as follows:

		Item			
		2	3	1	4
Machine	4	1	0	0	0
	1	1	1	0	0
	6	1	1	0	0
	2	0	0	0	1
	3	0	0	1	1
	5	0	0	1	1

The two cells can be formed as follows:

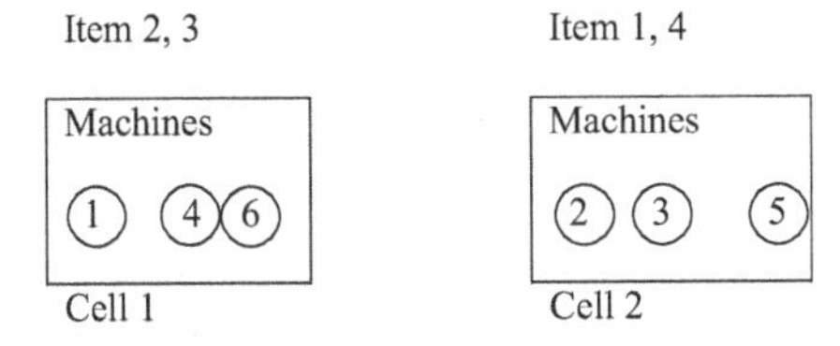

Machines 1, 4, and 6 are arranged in Cell 1, in which items 2 and 3 are processed. Machines 2, 3, and 5 are installed in Cell 2 for processing items 1 and 4.

Exercises

1. What are the basic flow lines for material flow? Describe their characteristic features.
2. Assembly line balancing is an important area of the analysis of product layout—comment.
3. Tasks from A–J are to be performed in an assembly line. There is a requirement of 400 products per day. Assume the productive time as eight hours per day. The task time and the immediately preceding tasks for each task are as follows:

Task	Immediately Preceding Tasks	Task Time (Seconds)
A	–	45
B	A	25
C	A	35
D	C	40
E	C	25
F	B	50
G	D,E	50
H	F	20
I	G, H	35
J	I	60

Balance the assembly line and determine the efficiency of your proposal.

4. Discuss the incremental utilization heuristic briefly. Using this heuristic, assign the tasks to workstations for the following precedence diagram of tasks. Time taken by each task is also provided in minutes.

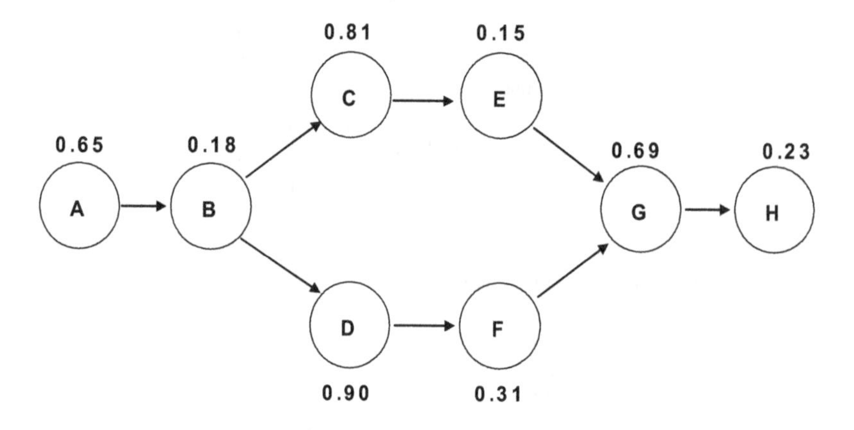

Assume that the assembly line will operate for 400 minutes per day and 2,000 products are required to be assembled each day. Also, evaluate the proposed solution by computing efficiency of the proposal.

5. Name the various techniques regarding planning for the process layouts.
6. How is systematic layout planning carried out?
7. Discuss various assumptions in the CRAFT.
8. Implement the CRAFT for the following initial layout.

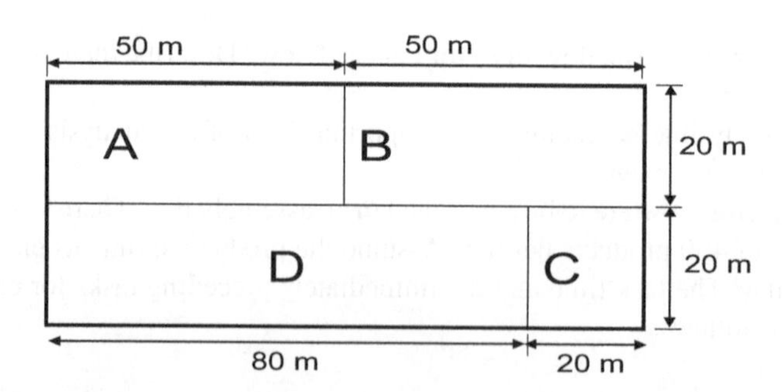

Load to be transported in certain period is shown in the movement matrix as follows:

From \ To	A	B	C	D
A		2	1	3
B	3		4	2
C	2	3		4
D	0	0	1	

Cost in rupees per metre of transportation is as follows:

From \ To	A	B	C	D
A		1	1	1
B	2		1	1
C	2	1		1
D	1	1	3	

11. How is the COFAD different from the CRAFT?
12. Discuss various closeness ratings in the ALDEP.
13. Six departments—*a*, *b*, *c*, *d*, *e* and *f*—are to be located in a layout of equal length and width, i.e. having square shape. The relationship chart is as follows:

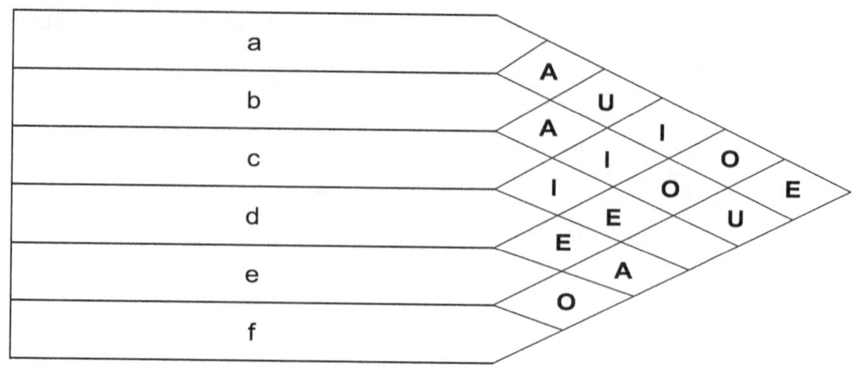

Departments

Area of each department is as follows:

Department:	*a*	*b*	*c*	*d*	*e*	*f*
Area in m2 (× 100):	6	18	20	6	20	30

Assuming the sweep width to be two columns, implement the ALDEP.

14. How is the CORELAP different from the ALDEP?
15. Implement the bond energy algorithm for the following machine-item details and comment whether the suitable cells can be formed.

Item

Machine		1	2	3	4	5	6
	1	1	0	1	0	1	1
	2	0	1	0	1	0	1
	3	0	1	0	1	0	1
	4	1	0	1	0	1	1
	5	1	0	1	0	1	0

Bibliography

Banerjee A., Sylla C., and Eiamkanchanalai, S, 1990, Input/output lot sizing in single stage batch production systems under constant demand. *Computers & Industrial Engineering*, 19(1–4), 37–41.

Buchan J. and Koenigsberg E., 1970, *Scientific inventory management*. Prentice-Hall.

Chase R.B., Aquilano N.J., and Jacobs F.R., 2000, *Production and operations management*. TMH.

Chowdhury M.R. and Sarker B.R., 2001, Manufacturing batch size and ordering policy for products with shelf lives. *International Journal of Production Research*1, 39(7), 1405–1426.

Golhar D.Y. and Sarker B.R., 1992, Economic manufacturing quantity in a just-in-time delivery system. *International Journal of Production Research*1, 30, 961–972.

Nahmias S., 2001, *Production and operations analysis*. McGraw-Hill.

Panneerselvam R., 1999, *Production and operations management*. PHI.

Pinedo M. and Chao X., 1999, *Operations scheduling with applications in manufacturing and services*. McGraw-Hill.

Plossl G.W., 1994, *Orlicky's material requirements planning*. McGraw-Hill.

Sarker B.R. and Babu P.S., 1993, Effect of production cost on shelf life. *International Journal of Production Research*1, 31(8), 1865–1872.

Sharma S., 2001, A fresh approach to MRP lot sizing. *Journal of the Institution of Engineers (India)-PR*, 81, 39–42.

Sharma S., 2003, Integrated lot sizing with temporary price discounts. *Industrial Engineering Journal*, 32(2), 23–26.

Sharma S., 2004, Optimal production policy with shelf life including shortages. *Journal of the Operational Research Society*, 55(8), 902–909.

Sharma S., 2008, On the flexibility of demand and production rate. *European Journal of Operational Research*, 190, 1557–1561.

Silver E.A., 1989, Shelf life consideration in a family production context. *International Journal of Production Research*1, 27(12), 2021–2026.

Silver E.A., 1990, Deliberately slowing down output in a family production context. *International Journal of Production Research*1, 28(1), 17–27.

Silver E.A., 1995, Dealing with a shelf life constraint in cyclic scheduling by adjusting both cycle time and production rate. *International Journal of Production Research*1, 33(3), 623–629.

Smith S.B., 1989, *Computer-based production and inventory control*. Prentice-Hall.

Tersine R.J. and Barman S., 1995, Economic purchasing strategies for temporary price discounts. *European Journal of Operational Research*, 80, 328–343.

Tersine R.J., Barman S., and Morris J.S., 1992, A composite EOQ model for situational decomposition. *Computers & Industrial Engineering*, 22(3), 283–295.

Tersine R.J. and Schwarzkopf A.B., 1989, Optimal stock replenishment strategies in response to temporary price reductions. *Journal of Business Logistics*, 10(2), 123–145.

Index